Best Practices in
Biotechnology Business Development

Best Practices in
Biotechnology Business Development

Yali Friedman, Ph.D., Editor

LOGOS
PRESS

Best Practices in
Biotechnology Business Development
Yali Friedman, Ph.D., Editor

Published in The United States of America
    by
Logos Press, Washington DC
www.Logos-Press.com
info@Logos-Press.com

Copyright © 2008 thinkBiotech LLC

All rights reserved. No part of this book may be reproduced or transmitted in any form or by any means, electronic or mechanical, including photocopying, recording or by any information storage and retrieval system, without written permission from the publisher, except for the inclusion of brief quotations in a review.

10 9 8 7 6 5 4 3 2 1

ISBN: 978-0-9734676-0-4

Library of Congress Cataloging-in-Publication Data

Best practices in biotechnology business development : valuation, licensing, cash flow, pharmacoeconomics, market selection, communication, and intellectual property / Yali Friedman, editor.
   p. cm.
  ISBN 978-0-9734676-0-4 (perfect bound)
  1. Biotechnology industries. 2. Intellectual property. I. Friedman, Yali.
  HD9999.B442B475 2008
  660.6068'1--dc22
                                    2008006115

Contents

Introduction — 1
Yali Friedman

Top Five Mistakes — 3
Carlos N. Velez

Communicating With Investors and the Media — 13
Fay Weston

Securing Intellectual Property From the Inside Out — 37
John Avellanet

Aggressive Intellectual Property Strategies — 61
Gene Rzucidlo and Stefan Miller

Market Based Business Development — 81
Ryan Bethencourt

The Ins and Outs of In- and Out-Licensing — 91
Gil Ben-Menachem

Free Cash Flow—the Essential Ingredient for Growing a Business — 111
Gerald S. "Sandy" Graham

Biotechnology Transfers and Models Facilitate Access to Biotechnological Inventions — 127
Oleksandr Skorokhod

Maximizing the Strategic Impact of Health- and Pharmacoeconomics in Biotechnology Companies — 141
Ulf Staginnus and Stephen Russell

Valuation of Technologies Through Services — 161
Ingrid Marchal-Gerez

How Big is the World Market for Biopharmaceuticals? — 175
Ronald A. Rader

Introduction
Yali Friedman

The biotechnology industry operates at the leading edges of science, business, regulatory, and political spheres. Biotechnology has impacted diverse fields such as drug development, agriculture, and industrial processes. The scope of possible applications is defined by scientific abilities, marketplace needs (and availability of investors), regulatory incentives and allowances, and political support. Innovations and developments in each of these areas appear daily, and are perpetually changing the shape of the industry.

The biotechnology industry is growing rapidly. Media reports regularly announce new biotechnology development initiatives supported by national and regional governments. Every year increasing numbers of life science graduates enter the workforce, more researchers seek to commercialize their research findings, and growing numbers of biotechnology-minded business students and business professionals seek to apply their skills and knowledge to biotechnology. Some of these biotechnology workers enter existing companies, whereas others start their own companies. Regardless of their education, training, or the nature of their work, this growing constituency is challenged to learn how to operate within the biotechnology industry, while continually monitoring and managing the implications of changes in the underpinning fundamentals.

Those operating within the industry—whether in biotechnology companies or in associated supportive roles—are constantly challenged to keep abreast of industry developments and understand their significance. There are many existing outlets which are very effective in communicating these changes and interpreting their significance, but I have seen a need for guidance to manage business development within this ever-changing industry.

Seeking to meet the need to understand how to practice the business of biotechnology, I set out to assemble a set of best practices to be used as a framework upon which to understand critical issues in biotechnology business development. I have selected experts from a wide range of disciplines and asked them compose best practices based on their experiences and expertise, with the intention of furnishing the reader with a vital toolbox covering a broad spectrum of topics. The objective of these chapters is not to enable the reader to, say, file patents or draft licensing agreements unassisted. Rather, it is my hope that by reading these best practices the reader will develop a better understanding of the key elements in these operations and become empowered to better manage their implementation.

I would like to close by thanking Harley King, a student in the MS/MBA Biotechnology joint degree program at Johns Hopkins University, who helped read and correct the entries.

Top Five Mistakes
Carlos N. Velez

Carlos N. Velez, Ph.D., MBA, has been involved in life science research, technology assessment, consulting, and valuation since 1990. He has substantial expertise in the assessment and valuation of life science technologies such as pharmaceuticals, diagnostics and medical devices. In 2005, Carlos leveraged this experience to form the Erie Hudson Group, a boutique consultancy focused on life science consulting, business development and financial advisory. The Group's clients range from multinational pharmaceutical companies to start up companies. In addition, the Group advises life science investors during the due diligence process. Carlos can be contacted at *carlosnvelez@gmail.com*.

The word *entrepreneur* frequently triggers thoughts of people ranging from a corner baker to Bill Gates (Microsoft), Richard Branson (Virgin), and Jeff Bezos (Amazon). The life sciences industry is no stranger to successful entrepreneurs, as shareholders of Alza (Alejandro Zaffaroni), Genentech (Boyer and Swanson) and Biogen (Phillip Sharp) can attest. Yet an entrepreneur can just as easily be the corporate executive who wishes to start a new business initiative, such as the pursuit of a new indication for a commercialized drug.

Regardless of an entrepreneur's background, many entrepreneurs fail to successfully accomplish their financial and professional goals. Why? Are there common pitfalls that many entrepreneurs fall into, irrespective of their background or business? What lessons can the entrepreneur learn from the industry executive turned entrepreneur, and vice versa?

This chapter focuses on the five most common entrepreneurial mistakes. While these mistakes (or lessons, if you prefer) are geared towards the traditional entrepreneur, they are just as applicable to "internal" entrepreneurs, or entrepreneurs seeking funding and resources within a corporation. In fact, I make no distinction between internal and external entrepreneurs for this chapter. We use the term internal entrepreneurs when discussing corporate entrepreneurs who seek funding and other resources in order to initiate and advance an opportunity. External entrepreneurs break away from the corporate or academic fold to start their own companies. Yet, both terms can be used interchangeably for purposes of this discussion.

MISTAKE #1 – NOT UNDERSTANDING THE RISKS

All entrepreneurs accept three basic types of risk. *Personal* risk refers to the highly personal risks an entrepreneur accepts. For example, entrepreneurs are frequently asked to make personal investments and sacrifices when initiating a new business. Family relationships are most susceptible to entrepreneurial vagaries, as many spouses are often asked to make sacrifices in family togetherness and lost income.

Financial risk is accepted either by investing personal funds in a new company (i.e., a second mortgage or home equity loan), or by shifting funds from one budgetary area to another. Note how this differs from the process of seeking external funding. An entrepreneur seeking external funding from a venture capitalist or other source is asking the investor to shoulder the financial risk.

Professional risk is accepted when an entrepreneur accepts a risk which can negatively affect his or her professional progression. Managers who repeatedly fail to lead initiatives within corporations often find themselves moving sideways or even declining in their professional development.

So what is the problem? Many entrepreneurs fail to take these risks into account before deciding to become entrepreneurs. For example, some entrepreneurs fail to discuss and carefully plan with their spouses before undertaking entrepreneurial activities. This can potentially result in extreme duress for both the business and family financial situations due to a lack of careful planning. Others fail to appreciate the financial or professional costs associated with a new venture, and hence do not create contingency plans for these risks.

MISTAKE #2 – TAKING THE WRONG FIRST STEP

Many entrepreneurs assume that receiving an investment (venture capital, increased budget, and so forth) is the first step before a company or initiative gets started. This is incorrect. Many entrepreneurs are unwilling to invest their own time and resources in the initial business-building steps like forming a limited liability corporation, or in-licensing a technology from a university in exchange for equity. Some entrepreneurs are reluctant to use personal funds to perform even simple tasks such as creating business cards, setting up web and email addresses, and establishing phone numbers.

The fact is that any entrepreneur, regardless if he or she is starting a biotech company or a bakery, will need to invest time and personal funds to get started. Today, it is extraordinarily easy to set up a website and obtain an email address, order high-quality, customized business cards, establish a business phone line, and so forth. Many attorneys and consultants, especially those working in geographies where start-ups are commonplace, may work for reduced billing rates in exchange for equity or other arrangements.

Similarly, many external entrepreneurs assume venture capital or angel investors are the only source of start up funding. If anything, with the growing scarcity of angel investors, local, state, and federal grant programs are becoming increasingly important for entrepreneurs. In fact, a cottage industry of independent grant writing consultants have emerged who specialize in writing SBIR/STTR grants. The Small Business Administration web site (www.sba.gov) is full of information on these programs. Also, a number of state governments, state universities, and even private universities have set up funds and other resources to assist entrepreneurs. New York is a prime example of a state that has a number of potentially helpful resources within the state government, state-university system, and within the major independent universities. Many of these programs provide entrepreneurs with cash, tax incentives, laboratory space, and other resources. Lastly, public-private partnerships are becoming increasingly willing to provide non-dilutive capital for innovative, high-risk projects.

MISTAKE #3 – MISUNDERSTANDINGS ABOUT THE FUNDING PROCESS

Many entrepreneurs do not have a sense of how long it takes to receive some form of equity funding. Assuming that venture capital is the right source of funding (and this decision is often questionable), entrepreneurs should plan on the process taking anywhere from 6 months to a year or longer to receive an investment. Look at it from the investor's perspective: the investor is evaluating multiple opportunities at once. All of them sound attractive initially. All of them have risks and challenges. The investor must decide which risks and challenges he or she is most willing to accept...all this while also being able to generate an above-average return on the investment for his *investor*.

This touches on a specific concern, which is that many entrepreneurs do not have an understanding of the venture capital process. Terms such as "IRR" and "Preferred Shares" are foreign to many scientists-turned-entrepreneurs. Many entrepreneurs do not realize a venture capitalist is managing the capital for their investors, such as pension funds and insurance companies. They do not understand the venture capitalist's responsibility is to return the capital entrusted to them, and also to generate above-average returns. Many entrepreneurs also have misunderstandings regarding equity dilution and control, and the role that softer issues play (i.e., management team personalities) in the analysis conducted by a venture capitalist. Lastly, there are entrepreneurs who still do not have a sense of what types of companies are appropriate for venture capital in terms of revenue potential, time to exit, value at exit, and so forth.

MISTAKE #4 – FAILING TO SEEK PROFESSIONAL HELP

Entrepreneurs frequently fail to seek professional help in creating their business plans and presentations. It is tragic to see great technologies and product concepts poorly packaged in mediocre business plans and text-heavy presentations. Without a high quality plan (especially a quality executive summary) and presentation, an entrepreneur will not even get their foot in the prospective investor's door! Many plans are too long (more than 30 pages), full of technical details (which are likely confidential, and hence should not be in the plan anyway), and emphasize the wrong points, such as excessive financial detail, which comes at the expense of under-describing markets, sales and marketing, and competition.

Worse still is a "business plan" that takes the form of a poorly conceived slide presentation instead of a well-written prose document, or disparate presentations which are not synchronized with the prose document. Having a professional prepare these documents and advise the entrepreneur on the process can save a lot of grief in the long run, and increase the probability of a successful financing.

There are some investors who will say that a full plan is not necessary. There are some who will say that ten slides or a three-page Executive Summary are sufficient for them to sign a check. For some prospective investors in selected sectors, this may indeed be the case. But, in general, undertaking a funding drive with ten slides is a grave mistake. The notion that one writes the full plan so that the investor can file it away and forget it is also a grave mistake. The business plan is first and foremost a document for the entrepreneur and the entrepreneurial team. It is an expression of how the company proposes to uniquely capture the economic opportunity that has been identified. Perhaps some investors will take the plan and file it away, but each sentence and each slide must contain a great deal of thinking and planning. This thinking and planning should be written down in the format of a business plan for the benefit of the team. In fact, the mere act of writing down a business plan will surface critical questions that should be addressed before presenting the plan to prospective investors.

Many entrepreneurs forget (or don't realize) that a business plan is indeed *a plan*. While the content is fairly standard, the plan is a uniquely crafted document which captures the thinking of the initial management team. The most important part of the plan is the steps describing how the entrepreneur and his team intend to capture the unique economic opportunity. It is easy to prepare a plan that describes a large and rapidly growing market like China, India, baby boomers, diabetes, cardiovascular disease, Google, Web 2.0, etc. Explaining to a prospective investor how this opportunity will be captured is entirely different.

The *how* question must be kept in the forefront during the planning process. How will the entrepreneur market this product or service? How will he or she retain and service customers after the sale? How will he or she approach potential licensing partners? The answers should be action-oriented, and this is what any prospective investor will want to know. Yes, the opportunity is large, but how will the entrepreneur capture it?

Much has been written about poor quality presentations. Many assume the use of bullet points on slides is the best way to present material because it is the default presentation design in slide programs. Entrepreneurs will have a difficult time generating interest and excitement among investors if presentations are delivered via 30-40 text-heavy slides. This is especially challenging for corporate entrepreneurs whose slide designs and formats are expected to be identical for all presentations across the organization.

An entrepreneur's objective is to sell the opportunity to prospective investors. Entrepreneur's must therefore create enthusiasm and momentum for their nascent business. Can one imagine a television show or commercial with a single, motionless camera? Or a newscast with the same background? Or a movie with a single actor? This is what one has when every slide looks the same. The message may be interesting, but an uninteresting "package" will cause audiences to lose interest quickly.

At the opposite end, some jubilant entrepreneurs go through great lengths to create very sophisticated slides with complex charts replete with, graphics, arrows, animations, clip art, and the like. They erroneously believe PowerPoint prowess equates investment attractiveness in the audience's mind! Who can listen and read at the same time with full comprehension?

Seeking professional help to prepare business plans and presentations mitigates these issues. This is especially true if the professional knows the industry and the investors in that particular industry space. The professional will know the information his contacts are looking for, and will not be locked into standard templates and formats in delivering the quality set of documents an entrepreneur needs.

There is also a second and arguably more important reason to hire a professional. Strategic issues can be discussed and debated with the professional before any plan writing. For example, what is the exact business model? Is this company generating targets and agreements with pharmaceutical and biotech companies? If so, how? Or is the company internally focused, keeping all discoveries for development and commercialization itself? Who is the best partner for development and what happens if the relationship sours? What therapeutic areas should be "given away"? Which ones should be retained? How will funds be used? How will cash flow be generated? Can and should we have our products developed and manufactured overseas?

More detail in these areas presents a clearer picture to prospective investors. Some parts may be speculative. After all, the business plan might say that the company will pursue an agreement with company X. But an agreement may not be in place at the time the plan is distributed. But that is not the point. The point is to have a clear road map in the plan with multiple options (e.g., if Company X declines, we will reach out to company Z). Good VCs understand that there will be ups and downs as the company evolves.

MISTAKE #5 – IGNORING THE REST OF THE WORLD

Every start-up company must think globally from inception. Focus should be placed in the following areas:

> Global Markets: How can the product or service be commercialized outside the home country? Even if the bulk of the opportunity is for domes-

tic markets, a solid international opportunity can make or break an investment opportunity.

Competition: Are alternative technologies available from India? China? Eastern Europe? Can competitors take advantage of talent or favorable economics? Are there intellectual properties or other concerns that can be addressed proactively?

Value Chain: It is increasingly common for companies to have portions of the value chain performed globally, from call centers to new product discovery, design, and development. How can an entrepreneurial initiative exploit these global opportunities?

The real danger in ignoring the rest of the world is that a comparable company thinking along these lines will gain the advantage, and, hence, a prospective investor will choose another opportunity.

BONUS MISTAKE #6 – CHOOSING THE WRONG PARTNERS

Another important mistake to avoid as an entrepreneur is the choice of partners. Too many times company teams are selected based on resumes, and not selected for long-term personal commitment to each other and the company. This is an extremely difficult issue. On one hand, companies need to place qualified, skilled people in key positions. Conversely, skilled but belligerent employees may not be able to work together. This situation is potentially more detrimental than hiring a friendlier, but less-qualified team. Investor and director experience can shore up deficiencies in the qualifications of an otherwise highly-functional and complimentary team.

THE SECRET TO BEING A SUCCESSFUL ENTREPRENEUR IS…

So what is the "secret" to attracting investors? What is the "template"? The bad news is that there is no template. However, the lack of a template gives entrepreneurs limitless freedom regarding the expression of their unique value proposition. The business plan and presentation are opportunities for passionate creativity. However, the lack of a template also highlights the need for entrepreneurs to seek professional help early in the process.

So, entrepreneurs must:

> Be prepared to assume some start-up costs, some of which may be significant.
>
> Consider all potential sources of funding or in-kind services, and ask for advice and networking opportunities.
>
> Plan on a lengthy process, and arrange personal finances accordingly. If a current employer can be involved in exchange for equity, then explore this option, but do not sign anything without legal counsel.
>
> Understand the process from the investor's perspective, and be prepared to speak their language.
>
> Understand what angel and venture capitalists look for in investments in terms of revenue potential and time to exit from the investment.
>
> Seek professional help in creating legal structures, business plans, presentations, and other materials to maximize the probability of success. Leverage these professional relationships in identifying

prospective investors and members of the management team. Do not underestimate the importance of having a sound product with a plan that has undergone multiple rounds of rigorous planning and development.

Communicating With Investors and the Media
Fay Weston

Fay Weston is director of Talk Biotech, an Australasian-focused communications consultancy dedicated to the life science industry and its investors. Fay began her career in the UK providing investor and media relations advice to European and American biotechnology companies before moving to Australia. She has advised companies on their preparation for financing rounds, stock exchange listings and follow-on transactions in the capacity of media and investor relations advisor. Fay can be contacted at *fayweston@talkbiotech.com.au*.

Communication is an essential part of any successful business. All companies, including biotech companies, come to a point where they need to employ a communications program in order to achieve and support their goals.

The benefits of a well-executed communications program include recognition of your company and its products or technologies among important groups such as potential partners and investors who will support your business in the tough times. A communications budget and plan should form part of your business plan. As with all plans, it will change as the company progresses.

Companies that don't communicate or don't communicate well either disappear off the radar of potential investors and partners or, in the worst case scenario, are punished by investors who feel let down by the company.

The communications matrix includes media relations and investor relations. The audience for media relations is clearly the media, and for investor relations is the investment community. However, the two do overlap and should be integrated to bring about the best results for your company. For example, positive stories about your company in the media can lead to greater attention from investors and a related upsurge in share price.

What Are Your Aims In Communicating?

Before you commence communicating with any audience you need to consider what impression you wish to make and why. At different stages of your company's development your aims in communicating with the media and investment community will vary.

For example, if you are an unlisted company, your communications with the investment community will be targeted at those financial institutions that can assist you in your next financing round or IPO. Your communications with the media will be aimed at raising your profile within the industry community. If your product is still in development your audience, in terms of media, will be more specialized. It will be the journalists who have an interest in young companies and new technologies and products. If you have a product that is very close to market or on market you may be communicating with journalists who write health stories that are relevant to your target patient market.

If you are a listed company you will have a wider audience in terms of investor relations. It will include existing and potential shareholders, analysts, brokers, and fund managers. Your audience might be sophisticated institutional investors or retail ("Mum and Dad") investors. You will need to update existing investors and tell them what the company is doing. You will need to introduce your company to potential investors and explain why what your company does is worthwhile and why your company is going to be a success. Your interaction with the media is likely to include the lay public to a greater extent and some of your interviews will be aimed at a non-scientific audience.

Your audiences will have varying levels of technical and scientific knowledge, they will have different levels of interest in your company, and your challenge is to successfully communicate with all of them and to both create and maintain their awareness of your business.

In order to make your company memorable, to both media and the investment community, you need to differentiate your company from the competition. Before commencing any communications, make sure you know what your company does, why it does it, how it does it, and why it is doing it better than any other company. This sounds obvious, but is often forgotten in an interview, and it is often the case that the company's key spokespeople have different ideas about the answers to these basic questions.

THE COMMUNICATIONS TOOLBOX

Key messages are a valuable tool in all communications, regardless of who the audience is. Developing a set of key messages should be one of the first things your company does before interacting with media and investors. A recommended approach is to start with a simple set of four or five questions and discuss these with senior management until a consensus is reached. Your key messages should cover the following:

- Corporate
- Product / Technology
- Stage of development
- Business model
- Finance

A typical set of key messages for a drug development company might look something like the following:

KEY MESSAGES:

Corporate
- Commercially driven company with ability to develop its products internally and through external collaborations
- Opportunity to bring multiple products to market quickly in major medical sectors
- Highly experienced management team with strong capabilities in overseeing global operations and research
- Management team supported by both an

experienced board and a world class scientific advisory board

Company Products
- A new class of drugs with broad applications in diseases of the [insert appropriate descriptor]
- X clinical trials ongoing, with Phase Y data showing safety & promising efficacy

Business development
- Company's product discovery platform generates multiple candidates for exclusive licensing opportunities
- Company will take products through clinical development until Phase II before seeking licensing partners

Finance
- Solid funding by leading U.S. and European industry investors
- X years of operating capital
- Low cash burn rate as a result of streamlined operations

Key messages will assist in keeping your communications consistent. All company spokespeople should be aware of them and they need to be reviewed regularly. For example, if you have recently done a licensing deal with a well known pharmaceutical company you might want to add that into your key messages as an example of how you do business. Your key messages should also encapsulate what your company's unique offering is. The example given above is generic, but the section on products or technology is the obvious place to start when trying to communicate why your company is unique.

Corporate Spokespeople

Your corporate spokespeople will set the tone for the company to the outside world. The company's reputation to a large extent will be based on their ability to communicate concisely and accurately about the company.

Consider nominating several spokespeople according to their areas of expertise if possible. It is unlikely one person will be able to answer all possible questions relating to the company. The main spokesperson should always be the chief executive officer. However, for communications about the science, products or technology, the chief scientific officer or chief medical officer may be more appropriate. Your board and chairman can also be used for communication when necessary.

Not everybody is a natural communicator and it takes practice. Spending some of your budget on presentation and media training is a worthwhile investment if your spokespeople are not experienced in this area. It pays dividends in the long run. You will not be expected to deliver your lines with a professional actor's panache, but the ability to communicate clearly is an absolute necessity.

Within the company, employees should be aware of who the nominated spokespeople are and should refer all inquiries appropriately. Many companies set up a standard operating procedure (SOP) to follow for all inquiries. This is particularly important if the company is running a clinical trial. Only those people qualified to answer questions and nominated to do so should respond to queries. It is obviously inappropriate for the finance officer to answer medical questions if he/she is not qualified to do so. However, if your company is publicly listed, your spokespeople also need to be aware of disclosure rules and how these will apply.

Materials

The typical materials for communicating include press releases, fact sheets that describe your company and technology, and the corporate website. There is no excuse for a company, whether it is a start-up or a more mature company, not to have a website.

Spending a lot of money on a website is not necessary. The most important thing is that you have an online presence, and that the information contained on your site is up to date. When people want to find out about your company this is where they will go. Your technology and products can be marketed on the site, and it is a useful place for your potential partners and investors to gain information about the company. It is also a valuable source of information for journalists.

If you have a website you already have a majority of the information needed to compile non-confidential fact sheets about your company. As with press releases on your website, it is a good idea to make these available as downloads. You may want to have a section of your website where viewers can download fact sheets about the company, the technology and products, disease information, and relevant images.

The fact sheets can be divided into a "corporate fact sheet" and a "product or technology fact sheet." The former gives key data about your company such as its history, area of expertise, field of operations, market data, key product or technology summaries, and management team. The latter gives more in-depth, non-confidential information about the company's products or technology. The aim of these documents is to give a brief introduction to your company and products. It is important not to get too detailed—if your audience is interested they will ask you for more information. It is also important the information is not too technical. Remember that your audience is not stupid, so don't patronize them, but at the same time, they may not be specialists in your area, so avoid jargon and unexplained acronyms in these documents.

A corporate kit does not have to cost a fortune to produce, and in the early days of a company's life it is still possible to have a professional looking corporate kit without breaking the budget. The corporate kit should contain the business card of the chief executive officer or relevant contact, a corporate fact sheet, a product/technology fact sheet, recent press releases, and any particularly good media articles or reports about your company. All of these items can be produced using Microsoft Office if budget is an issue. For media excerpts and reports you will need to check on copyright restrictions. These kits should be a useful source of information for media and investors and are ideal to leave behind after meetings or use at conferences.

COMMUNICATING WITH INVESTORS

Whether you are a private or a public company you have investors. They may be venture capitalists, a university, family and banks, or they may be financial institutions and retail investors. These people, while they are not taking care of the day-to-day running of your company, are the mainstay of your business. It is almost certainly the case that without their support you would not be in business at all. As such, you have a duty to keep them informed about what you are doing with their funds.

What sort of information does a typical investor want from you? The answer to this question varies depending on the investor, but you can be sure they want to know you are doing something with the money they have entrusted to your company. If they expected you to do nothing with it, they would have invested it in a savings account, not your company. Most of your investors will not want a blow-by-blow account of every development in minute detail. What they will want to know is how their investment is being spent, and whether these expenditures are bringing benefits. The very fact that you have investors means that at some point you have written a business plan, and your investors have seen this or something similar. They will measure your success on the basis of this information and this is a good guideline to use in your communications with them. As you continue dialogues with your investors, keep notes on what you have communicated and refer back to this before you next communicate with them. Your company is likely to be one of many that your investors are communicating with, so recap what the business has been doing recently and relate your current topic back to previous discussions and disclosures.

There are numerous ways of communicating with your investors, and in today's global economy your investors may be geographically dispersed. It is crucial that you don't alienate your existing investors by failing to communicate with them. For example, company X, based in California and listed on the NASDAQ, has a majority of its investors based in the USA, but also has a significant number of investors based in Germany. In order to keep all of the company investors informed, Company X does several things:

Firstly, all of the company's documentation is available in both languages. When disclosures are made via press release on the NASDAQ they are also translated into the German language and disseminated across the German wires to media and investment houses in Germany. Investor briefings are done via videoconferencing with suitable sub-titles if necessary, and are either arranged at times suitable for both countries or are disseminated via the web. The management team of company X also ensures that its German shareholders have the opportunity to make face to face contact with the company at least once a year.

It is always worth getting to know your investors, and gathering investor related intelligence is very valuable. If you know the style and focus of a particular investor you can plan for likely questions and prepare for your meetings with them with more confidence. For example, if you know a particular investor has a penchant for immunology companies (assuming you are also in the immunology space) you should familiarize yourself with your competitors and be prepared to differentiate yourself in these terms.

While your communications should always be consistent, you can tailor the focus to suit your purpose. One presentation does not suit all. If you are dealing with a general investor who is not a specialist in the life sciences area, giving a presentation with vast amounts of clinical data is not appropriate.

Taking the time to research your potential investors pays dividends not only in terms of success in getting a meeting or securing the investment, but also in terms of time management. If, having done some research on a potential investor, you discover that they only invest for the short term, you will want to bear this in mind when seeking a meeting. What can you offer them in the short term and do you actually want this type of investor on your share register?

Building a relationship with your current and potential investors takes time and you will rarely see results overnight. Potential investors may not be able to invest in your company right now, but keep them informed because things change quickly and a glut of investors is an almost unheard of phenomenon in biotech. These relationships need to be nurtured and they will contribute to the company's future success.

Disclosure

As a private company you enjoy relative freedom in terms of how you communicate with the public and your investors. This all changes when you become a public company by listing on a stock exchange. However, it is never too early to start a policy of "best practice" in your communications.

There are a number of laws that relate to disclosure in each country, and all stock markets have regulations that govern the disclosure of sensitive information. Examples include the US Securities and Exchange Commission's Regulation Fair Disclosure, the Australian Securities & Investments Commission, and Australian Stock Exchange's principles of continuous disclosure, and the UK Financial Services Authority disclosure rules. The aims of these regulations can be summed up in the phrase "non-selective and timely" disclosure to investors. Rather than going into detail on any particular geographical incarnation of these rules, what follows is a broad brush picture of "best practice."

The key is in understanding what constitutes "material" or "price-sensitive" information. A simple way to think about this is to ask yourself, "If I knew about this and I was an investor, would I trade on the basis of this information?" If the answer is "yes" then it is price sensitive information and must be disclosed as soon as is practical. At the same time you should not disclose incomplete information that may unduly influence the market in one direction or another. An example of the latter would be issuing a press release claiming a successful clinical trial result but failing to disclose that a safety endpoint was not met.

Best practice disclosure includes both good and bad news, and delaying or failing to disclose what might be interpreted as bad news is not recommended. An example would be that you discover the clinical trial for your lead product is having recruitment problems and this is going to delay the results. The market believes you are going to be in a position to announce results at the end of the year because you stated that it was a 9 month trial when it commenced. This news may have a negative effect on your share price, so what is the advantage of disclosing you will not meet this milestone earlier rather than later?

The advantage is that by letting your investors know in a planned manner as soon as you have sufficient data, you are protecting your company's reputation. You put the company in a position where investors may ask questions about how long you have known this and then question the trustworthiness of your communications by waiting until you are asked about missing a projected milestone. If there is a good reason why you will not achieve a milestone, tell your investors about it ahead of time – they will appreciate your candor and honesty, even if your share price suffers in the short term.

For the same reason, if you become aware of a rumor that could effect how your stock trades, deal with it sooner rather than later. You cannot prevent speculative trading, but you can minimize the effect it has on your company in the long term. Well informed investors are more likely to support you in the long term. Investors who feel ill-informed or misled will, at the very least, exit their investment to your disadvantage.

A good rule to live by is "Always under-promise and over-deliver." Remember that you will be held accountable by your investors for everything that you say and that delays are common. For example, don't give fixed dates on events such as the close of a clinical trial or the signing of a deal — these are all events that can be delayed through no fault of the company. Your communications need to be transparent but that does not mean that you have to answer every question posed if it might be to the detriment of the company. For example, if an investor or a journalist asks you when you are going to sign a licensing deal and who your intended partner is, you are under no obligation to give them that information until the deal is done, and to do so could jeopardize your current negotiations. You might respond by saying "At this stage we are in negotiations but I am not at liberty to disclose who with. I would hope to be able to update you in the next couple of months/weeks."

COMMUNICATING WITH THE MEDIA

Communicating your company's story to the media can be challenging. There are various types of media including the financial press, the trade/specialist publications, the more general press, and they all require different approaches. One thing they all have in common is a hunger for news and deadlines.

For some executives, the most difficult moments of their careers occur when they must face a reporter, camera, microphone, or take a reporter's call. Try to be positive about the media. Reporters have a job to do and almost invariably try to do it honestly. Reporters are professionals representing the public and they feel the public has a right to know the facts. A reporter or writer has one principal objective: to find out as much information as possible about your company (or a particular product or issue involving your company), and to publish/broadcast it before their competitors. Keep in mind that reporters are professional information gatherers. They can take a lot of seemingly innocuous pieces of information and put them together to come up with a story.

Whether you have instigated an interview or the reporter has come to you, preparation is essential. It is important to look at an interview as a positive situation and an opportunity to secure positive exposure. Interviews are an opportunity to promote your company and get free publicity. By following a few simple guidelines such as preparing in advance, establishing credibility, and communicating something newsworthy, these potentially difficult situations can be turned into successful encounters.

Interview Preparations

Time spent preparing for an interview is never wasted. You should gather information about the reporter such as what he/she has written in the past and what the purpose of the interview is from their perspective. Ask the reporter beforehand what aspect of the subject he is interested in. The purpose of this is to get a feel for how the reporter thinks and what questions he/she might ask you. Also make sure that you are familiar with the publication or radio/TV program itself—you don't want to appear ignorant or insult the reporter by not knowing it. When you are researching the publication, think about what the publication's audience wants to hear and what they want to find out.

Based on your research you can target your communications to satisfy the type of publication and audience that will be reading it. You should consider the points that you want to make and think about what your goal is in the interview. As part of your preparation you should know the one, two, or three key points you want to make and have simple facts and figures ready to support those points. It often helps to list the positive things you want to talk about and then consider that if you only had time to say three things what they would be?

Prepare something substantive remembering reporters are interested in "who, what, when, where, how, and why." In preparing answers to possible questions, you should make every effort to secure strong substantiating information—dates, facts and statistics. These all lend authority and credence to your answers. A reply based on empirical findings always holds more weight than one's opinion.

If you want to appear in major media, you must say something that will interest the reporter and that is newsworthy. It is newsworthy if it makes sense and is being said for the first time. You must have a unique and strongly held point of view. Prior to all interviews, it is often useful to prepare a single sentence, a catchy phrase or brief analogy that summarizes your views on an issue. The media almost universally appreciates quick summaries. They will look for them, and if good, will use them. The reverse is also true: a long, rambling, disjointed response with long pauses between awkward clauses will practically ensure that no part of the interview will be used. Answers must be quotable, not esoteric, reflective or convoluted. If the information bores a reporter, he or she will think twice before contacting you in the future.

A vital part of your preparation is ensuring you do not breach any disclosure regulations and keep only to information that is already in the public domain.

Interview Tactics

Your first task in an interview is to establish credibility. Demonstrating respect for the reporter, appearing open, honest and responsive to the reporter's questions are qualities that will improve your relationship with the reporter. If you lose control of the topic, you lose credibility as an authoritative, knowledgeable speaker. If you appear on TV and have only ninety seconds to talk, do you want to simply answer the reporter's questions or do you want to get *your* message across? Always answer the questions, but the real skill is to say what you want to say irrespective of the questions.

Use your message points at the beginning of an interview and don't wait for opportunities as the interview will be over before you know it. You control the interview and your answers often determine the next questions. Surprisingly, you can often prompt a reporter to shift his/her focus by saying something like, "Those are good questions, but did you consider asking about …, which may actually be more central to this matter than it would seem at first glance?"

This technique is called "bridging" and can be utilized to answer the questions you want to answer (not necessarily the ones being asked). After briefly responding to the question posed, phrases such as: and, but, however, on the other hand, can all provide segues back to your message points.

You should always aim to give your answers in the positive. For example, rather then stating that you have an increased loss this quarter, say, "Our revenues, including product sales have tripled this quarter." If an interviewer phrases a question or statement negatively, do not repeat it. The question should be acknowledged, but followed by a positive action to put the subject into perspective, for instance, "You are right by saying it has been a problem, but as our recent figures show, we are still the leading company in…" Often, a positive attitude and enthusiasm will create a good interview.

The reporter will never know as much as you about your subject but he or she may think he/she does. Be confident, you are the expert on the subject. During the interview remember the "Three C's"—Concise, Conversational and Catchy. Typically, your comments will be edited to about 5 to 15 seconds or a short sentence. Remember, a reporter's goal is to fill column space (or fitting information into a limited amount of column space) rather than filling time. Focus on getting your points across efficiently and avoid long words and lengthy sentences. Also, it is better to pause to gather your thoughts than to rely on fillers like "uh-uh-uh," "like," or "you know." You should avoid insider jargon and policy-laden language and use words and descriptions that the average reader/viewer will understand. When you must use jargon, explain it—briefly. The reporter is looking for the catchy phrase or sound bite. To ensure your main points are included, say them in a clever fashion. If you just presented a key point in an unclear or rambling way, stop for a second and make your point again. The reporter needs the quote to make sense.

Your opening and closing statements are often the most critical. Remember especially to reiterate the three most important points again in your closing statements. Be clear and concise. It is often the last word of the interview that is taken away with the reporter.

"Off the record"

This is used to broaden the journalists' knowledge about the subject and get a better represented article. It must be stated that something is "off the record" before it has been spoken. It is sometimes acceptable for a quote to be used as long as it remains noncontributable. While reporters are supposed to respect the confidentiality of any statement you give "off the record," in reality, there is no absolute guarantee the reporter won't use it. If you do not want something printed or broadcast, do not say it. If the reporter is likely to quote you, then you can ask him to read back your quotes after the interview. The safest assumption to make is that there is no such thing as "off the record."

Interview Do's & Don'ts

- Do not allow the reporter to interrupt: Be polite but firm. Ask the reporter to allow you to give full answers.
- Do not let the reporter put words in your mouth: Some reporters summarize what you have said—or what they think you have said. Don't be afraid to repeat yourself or restate your message in several ways if you feel the reporter does not understand you.
- Don't fill awkward or silent pauses: The reporter may pause to try and make you continue speaking and volunteer extra information. If you have said everything you wanted to, say so or continue with what you want to talk about so you end up taking control of the conversation.
- Answer questions concisely and directly: Choose three main points that you want to say and make sure, if you say nothing else, that these points are mentioned. Do not get involved in a discussion that may lead you into quicksand, causing you to say things you do not want to discuss. Once you have completed an answer, end it there. Don't waffle.
- Never lose your calmness: Act positive, businesslike and forthright. Don't give an angry or defensive response—simply redirect your answer to one of your main points. Reporters get the last word, so don't get into a verbal tussle.

- Don't repeat the reporter's words: At times a reporter may use language in a question that is confusing or even negative. The question won't appear in the final version, but your answer will, so don't repeat it.
- Never repeat an incorrect statement made by the reporter: You may be quoted, and if your statement is taken out of context it will appear that you accept or agree with the statement.
- If you don't know the answer, say so: Offer to find out the information and get it to the reporter as soon as possible. Never respond to a question with "no comment." It sounds like you're hiding something. Rather, generously describe why you cannot specifically answer that question and direct the conversation back to one of your main points.
- Don't be led into hypothetical situations: A media favorite is the "what if" question. If the reporter says, "Assume that…" or "What if…" and you don't like the direction being taken, respond with something like: "I can't speculate on the unknown, however…" and restate one of your main points. You don't have to play this dangerous game.

Not all interviews result in media coverage, but when they do, take the opportunity to review the results. If you are satisfied with a story, let the reporter know. A short thank you note to the reporter stating you appreciated the opportunity to talk with him or her and that you very much enjoyed the article is an important touch in establishing a long-term relationship.

CRISIS MANAGEMENT

Let's assume your company is doing well and your communications programs are running smoothly. And then disaster strikes. In answer to the question "What should you do if you found the wind was blowing your ship onto a lee shore?" The 200 year old Royal Navy Manual of Seamanship states: "DO NOT FIND YOURSELF IN THIS POSITION!" This is sage advice, but does not help if you do find yourself in that position. The key to weathering a crisis is planning. Crisis communication planning is essential no matter how big or small your company.

To plan your communications strategy for a crisis situation you need to have an idea for what exactly will constitute a crisis for your company. Examples might be:

- Product problems including clinical trial failure for the lead product, key technology failure
- Manufacturing glitch for product leading to lack of supply or a product recall
- Death or severe side effects in a clinical trial
- Financial issues such as share crash, hostile takeover bids or fraud

For many people, when something goes wrong in a major way, the instinct is not to talk at all or to talk too much. Neither of these responses is appropriate in a crisis time. The company's key responsibility is to communicate in a timely manner to all of its stakeholders. It is of paramount importance the company reacts immediately to contain the crisis, maintains timely control of information, speaks with a unified voice, prevents and diminishes the spread of rumors, maintains contact with external audiences, and protects the company's long term reputation and financial security.

Prior to a crisis event, nominate a crisis response team. This will usually consist of the chief executive officer, chief financial officer, senior vice presidents and representatives from regulatory, operations, marketing, human resources, media/investor relations, and legal. Specific roles for crisis team members should include a team leader who will initiate and coordinate the plan, spokespeople who will disseminate corporate messages, a logistics coordinator to coordinate conference calls, message dissemination, press announcements, and crisis communication plan implementation, and advisors on communications, financial matters and legal matters.

You will need to develop a tactics outline or procedures protocol which includes priorities and procedures for potential needs such as a duty roster, compilation of contact numbers for the crisis team, obtaining full information about the crisis situation, and logistical factors such as telephone and videoconferencing services if these are needed.

In a crisis situation you need to identify your primary audiences and your spokespeople. Your primary audiences may include the media, the financial and investor community, regulatory bodies, medical agencies, patients, advocacy groups, employees, customers, local authorities and partners. There should be procedures in place for disseminating information to these audiences. The chosen spokespeople should be appropriate for each target audience. It is also important to identify and brief external collaborators, industry experts and/or clinical site investigators who may be called by media, financial and medical communities for comment.

In terms of media you should have a complete up-to-date list of key reporters who regularly follow your company. You should also have an up to date corporate kit and a working Q&A document, positioning statement and a set of press release drafts for such an event. Consistency is the key.

If a crisis situation occurs how do you react? The following guidance gives an indication of how your planning might mitigate a crisis.

DURING A CRISIS:
Assess the situation
- Obtain full information about the crisis situation

Debrief crisis team members and corporate spokespeople

Review all procedures
- Prioritize sequence of activities
- Maintain control to allow swift activation of plan

Notify key audiences
- Determine appropriate approach for each audience
- Employees and external collaborators

Establish a "Command Post"
- Nominate a crisis command post to provide a central point from which all information flows and where the crisis management team can meet to obtain updates
- Set up a protocol for receiving any inquiries and provide instruction to anyone who may answer phones

Conduct media relations
- Develop appropriate messages based on situation
- Update all relevant background/Q&A information to ensure accuracy and uniformity
- Determine means for message dissemination
- Train spokespeople in advance to respond to incoming inquiries
- Respond to media calls on a case-by-case basis
- Monitor ongoing press activity related to crisis

Track public opinion
- Track reporting and public presentation of information as crisis proceeds
- Show that the company cares for those affected
- Maintain a good working relationship with the media
- Develop and maintain telephone log of all inquiries
- If a partner is involved, consult them before making any statements
- Adapt messages, as necessary, and correct misquotes or misperceptions

- Complete a press clipping portfolio

Caution — Don't:
- Admit legal liability unless specifically empowered to do so
- Lie or try to hide behind "no comment"
- Blame anyone or anything

Evaluate the results and update your crisis plan when the crisis is over. Depending on the nature of the situation, you may have been able to avoid going into full crisis mode. If this was not possible you will have conducted a valuable operation in terms of damage minimization, and hopefully your company's reputation is still intact.

MANAGING COMMUNICATIONS

If you are a small or medium-sized company how do you manage all of the above? The time requirement cannot be underestimated. As a result, most companies outsource this work to consultants until they are large enough to bring in a dedicated communications manager.

Finding a consultant to assist you with your communications can be a challenge. But, as the biotech industry grows, so do the number of service providers supporting it. There are a number of excellent media and investor relations consultancies specializing in biotech. It is definitely worth finding a specialist, as these people will have a working knowledge of the appropriate media and investor contacts, and will appreciate the specific challenges that face the biotech sector.

It's beneficial to find one consultancy that will handle both media and investor relations for you where possible. This saves on duplication and gives your consultants a much broader understanding of your business. There is no reason to split the two functions, as both elements are essential to your overall communications plan particularly in the early stages of your company's development.

The most important thing management can do in terms of

communication with media and investors is to set aside the time and budget to make it a fixed part of corporate strategy and planning. In order to obtain the best results, the communications program needs to be closely aligned with the business plan. This allows the company to develop the most appropriate communications strategies and take a more proactive and objective approach to its interactions with media and investors.

Finally, there is the issue of measuring the effectiveness and success of your communications program. How can this be quantified? One approach is to set out some key objectives at the beginning of the program. These need to be realistic – if you are a relatively unknown small company with early stage products when your communications program commences, you should not expect to be appearing on the front page of the *Wall Street Journal* and *Financial Times,* or to be entertaining large numbers of prestigious investors who want to plough money into your company in twelve months time.

A broad objective might be to increase awareness of your company among a key set of trade journalists and potential investors or analysts. More specific objectives might include the following:

- Obtain media coverage on a specific topic or a story on the CEO and company in a nominated publication
- Obtain X number of interviews with target journalists on the basis of forthcoming press releases
- Obtain meetings with X number of analysts/fund managers to introduce your company

Specific objectives can be easily measured. When appraising your activities remember that meeting the objectives can be confounded by a number of factors that don't necessarily reflect mistaken judgments in the communications strategy or failures in tactical execution. The most effective communications strategies are regularly reviewed and tailored to the changing circumstances of the company itself and its audiences.

In broader terms you might start by conducting a small scale perception audit before commencing your communications program. Identify a small number of journalists and members of the investment community who would be considered your target audience and develop a short set of questions to ascertain what they know about your company and their perceptions of the business.

Do they recognize the company name, do they know what the company does, can they name any of your technologies or products or give you key information about these, and can they name the key members of the management team? These questions will help you gauge the existing level of recognition and awareness surrounding your company. If there is base level recognition, you may want to ask for opinions on perceived strengths and weaknesses of the company, perceived challenges for the company and barriers to investment. The data will be invaluable in assisting with the development of your key messages and refinement of the communications strategy.

Repeat the exercise 12 months later using the same questions and measure the results against the original audit. You should see improvements unless your communications work over the past 12 months has totally missed the mark. You may discover a perceived challenge to the business has now been addressed or a perceived weakness has arisen in the course of the year due to changes in the industry. If a lack of understanding or misperceptions about the business persists, you need to revisit your strategy urgently and ascertain why it is not working.

It is likely some elements of your messages are being picked up, but others (usually the more technically-oriented ones) remain problematic with audiences struggling to understand the context. Common problems are the key messages not being adequately represented or the story is too complex for the audience. Further analysis may reveal that due to a lack of latent knowledge about the applications of a technology or the relevance of a novel therapy, your messages cannot be communicated easily. If this is the case you need to spend more time educating your audience and simplifying your messages.

Communicating effectively about science, technology, the biotech industry and business strategies that usually involve extended periods with limited revenue and considerable patience on the part of your investors, has never been easy. However, it is achievable, and as a biotech executive, it is a challenge that you cannot afford to shirk if your company is to become the next Amgen.

Securing Intellectual Property From the Inside Out
John Avellanet

John Avellanet is an internationally syndicated author, public speaker and business compliance advisor recognized world-wide for his expertise in FDA regulation and product development environments. He is a repeat guest on business radio shows such as Tomorrow's Business and My Technology Lawyer. Currently he serves as the managing director and principal at Cerulean Associates LLC, the consulting firm he cofounded. John can be contacted at *john@ceruleanllc.com*.

Intellectual property is the greatest asset of any company; it must be thoroughly protected and secured.

Companies that rely upon patents, trademarks and paper pledges of confidentiality are like home owners relying upon serial numbers, trespass signs, and product registrations to thwart would-be robbers. Such measures keep the honest from temptation but do little to ward off those with darker intent. In the knowledge marketplace of the 21st century information is the currency that buys competitiveness. Firms that lose control of their intellectual property (IP) will spend far more of their bottom line defending their IP in the courtroom than they would have if they had elected to secure it appropriately in the first place.

This chapter provides eight strategies for safeguarding intellectual property. These tactics are used around the world by firms large and small, and are appropriate for both inside a company, in its relationships with employees and contract personnel, and outside a company, as it deals with outsourced providers, vendors and marketplace development partners.

The advice mentioned here allows the biotechnology firm to keep its IP to itself, sharing components of its IP where, when, how and with whom the firm wants. It provides long-term flexibility for licensing, partnering and further innovation development. The firm, however, must balance its other needs with the recommendations herein – not all of the tactics listed here are viable for an eight-person startup, just as some of the tactics will require considerable effort in an 8,000-person organization. Whenever appropriate, text boxes present alternative approaches for the small startup as well as suggestions for further resources specific to the topic at hand.

The recommendations herein can also be useful to the potential investor or partner: if the biotechnology firm has not taken these (or similar) steps to secure its most valuable asset, then how likely will it be to guard your investment?

The eight rules for securing intellectual property are as follows:

1. Leverage initial legal advice
2. Secure an information framework
3. Balance access with isolation
4. Appoint an inventor "gatekeeper"
5. Accept the two changes of personnel
6. Craft a development agreement
7. Summarize IP twice
8. Circumvent your own limitations

LEVERAGE INITIAL LEGAL ADVICE

Lawyers protect their client's interests through available means in the legal system. In the case of intellectual property, this means patents, trademarks, service marks, copyrights and trade secret designations. But these designations simply serve as a clear demarcation of a firm's property; they do not secure it from theft or abuse.

Therefore, seeking legal advice is only the beginning step a firm must undertake to protect its information-based assets. First among the activities to be accomplished is disseminating the counsel's advice amongst company personnel, from the chief executive to the summer intern. However, because legal advice typically has multiple layers of meaning based upon a listener's experience, expectations and responsibilities, the company must create a program designed to allow the legal recommendations to be easily understood, adopted and put to into use by a wide audience.

To best leverage legal counsel's advice, design an intellectual property awareness program with at least three core components:

a. A cross-functional IP security team
b. An intellectual property primer
c. An IP liaison

THE CROSS-FUNCTIONAL TEAM

At the same time a biotechnology company's executives are engaging counsel (whether an internal individual or an outside firm), they must also reach out to the other members of the company accountable for protecting information property and new product development. At a minimum, this includes the information technology (IT) department (or in the case of a start-up, the

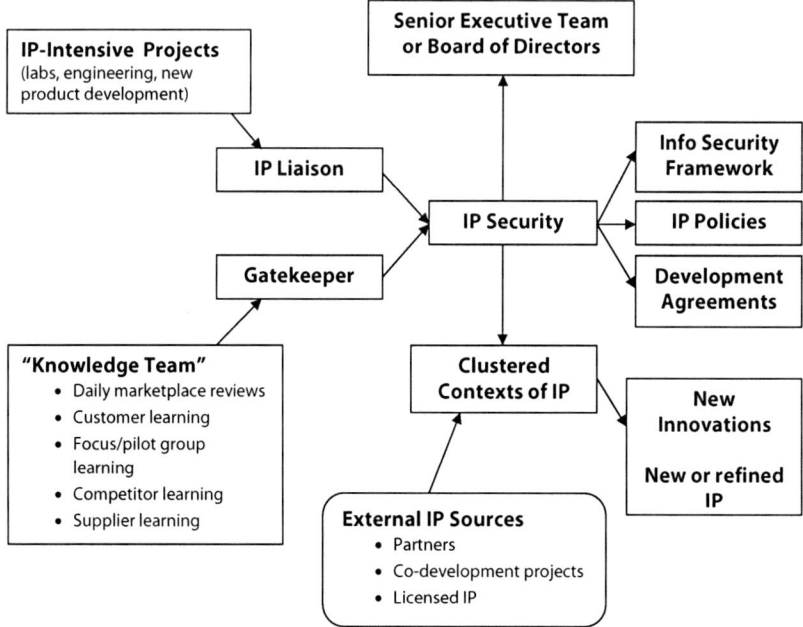

Figure 1: Inside-out intellectual property model

individual accountable for computers and software).

As a best practice, this includes individuals in the company who are responsible for regulatory compliance—from the Food and Drug Administration (FDA) to the Drug Enforcement Agency (DEA), from the Bureau of Industry & Security (BIS) to the Occupational Safety & Health Administration (OSHA).Those accountable for creating the intellectual property in the first place, the scientists and the engineers, and those individuals or departments who seek to sell the innovations and new products such as marketing and business development, should also be included..

Bring together a management-level individual from each of the above areas into a cross-functional team facilitates intellectual property security. If the company has an in-house legal department, select one of the lawyers as chair of the team, charged with leading and guiding discussions. Have the team report to the company's senior team or the board of directors.

There are two benefits of spreading IP security accountabilities across an organization: with many eyes watching, the risk of a lone individual causing harm is significantly reduced while the ability to proactively identify, and guard against, future risks is enhanced.

The Intellectual Property Primer

An essential, and often overlooked, component of effectively leveraging legal advice is working with counsel to create an intellectual property "primer." The primer must be crafted for a general audience (consider using Flesch-Kincaid readability tests to measure the difficultly of the primer's language; target a reading ease score of 30-60 or a grade level between 10-12). This is not an academic study of intellectual property history and rules, but rather, a high-level overview of the law that draws explicit connections between intellectual property and financial results in a knowledge-based economy.

The primer not only provides the background logic to all strategies and tactics the company implements for protecting its intellectual property, it also enables the company's personnel to make

> ## Small Startups and the IP Primer
>
> Most startup firms do not have significant resources to devote to legal counsel, and as such, should give serious thought to using publicly available resources on intellectual property written for a general audience in order to provide themselves a low cost primer on intellectual property.
>
> One such example is Nolo Press and its publications and article reprints. Be cognizant, however, that publicly available information should not be used in lieu of legal advice tailored specifically to the company or its intellectual property assets.

wise decisions in the field and in the lab. Therefore, structure the primer according to the following:

1. Layout the basics of intellectual property law.
2. Review the differences in the various IP designations (trademark, patent, trade secret, etc.).
3. Identify—and provide examples of—typical components of IP, from formulations to drawing schematics and unique processes.
4. Review the logic of the inventor/creator as owner and the role of his/her employer.
5. Give readers or presentation attendees typical ground rules to help identify when they might need to ask for advice.
6. Provide multiple real-world examples tailored to the company at hand, or the audience's role within the company; these examples should allow for an interactive level of participation to test for comprehension and the know-how to easily apply the rules day-to-day.

Finally, consider reviewing the company's approach, philosophy and policies in regards to intellectual property. Depending on the audience—employees, on-site contractors, vendors and third-party partners—there may be several layers of detail to cover. Identify the members of the cross-functional IP Security Team as the first level of help available for IP questions. While these individuals will not provide legal advice, they will screen concerns and direct questions to the appropriate resource, be that the IT department, legal counsel, regulatory affairs and so forth.

THE IP LIAISON

From the cross-functional IP Security Team, select an individual capable of understanding both the legal and the business consequences of treating something as intellectual property, including its disclosure (such as in a research paper); this individual will be required to make rapid judgment calls, often during product development meetings, and will need to be easily accessible to the scientists, engineers and lab managers. Ideally, such an individual will be discerning and perceptive; IP is not leaked or stolen from computers or file folders labeled "intellectual property," but gained from people.

The Intellectual Property Liaison will need to receive a few days of detailed training on IP, confidential information rules, trade secrets and patent disclosures by the company's counsel. Some larger organizations provide this detailed training to the entire cross-functional IP Security Team so as to ensure an IP liaison for each functional area of the company; for this section's purposes, the IP Liaison is directly associated with the product development, research and development (R&D) areas of a company.

The IP Liaison has nine core accountabilities:

1. Reviewing disclosure / confidentiality / copyright notices
2. Reviewing new product-related documents and presentations for appropriate disclosures and confidentiality / copyright notices (include both internal and external presentations and documentation)
3. Project summaries

4. Highlighting the statuses, risks, costs and pros and cons of any new product development projects, statuses, risks, costs and pros and cons
5. Monitoring IP information integrity and security
6. Working with the biotechnology company's IT department (internal or outsourced) to assign and review electronic security permissions to documents, files, drawings and so forth
7. Inventorying and tracking IP information
8. Comparing the roles and accountabilities of each individual with access to IP components
9. Communicating meeting summaries to appropriate company executives

Underlying this accountability is the essential ability to proactively raise a red flag for potential industrial espionage—"Why is Stuart requesting access to the engineering drawings when he is the analytical lab manager?"

IP-related meeting summaries are not minutes, but rather one to three paragraphs summarizing the discussion of any IP-specific subjects, decisions and revised development product timelines. Incorporate information pointing to or from patent applications, demonstrating awareness of and consideration for compliance risks (including patient safety, drug efficacy and so on), and identify any upcoming personnel changes.

Some companies find it helpful to assign a full-time IP Liaison to each new product project; other companies have been forced through necessity and resource limitations to create a shared role. Regardless of the approach taken by the biotechnology company, the IP Liaison must have the flexibility, and the maturity, to meet with both senior executives and the company's governance board to review any concerns he/she has regarding the new product project, the individuals involved or other items specifically related to protecting intellectual property.

Treat discussions with the IP Liaison as confidential. Patterns within the information and assessments presented by the IP Liaison may be subtle and require accumulation before they emerge.

SECURE AN INFORMATION FRAMEWORK

The IP Liaison (or the chair of the cross-functional IP Security Team) can also take the lead in developing an information security framework around the company's IP assets by following these three steps:

    a. Characterize the IP
    b. Assess risks and priorities
    c. Identify and implement controls

CHARACTERIZE THE INTELLECTUAL PROPERTY

Information assets cannot be compared to typical capital assets like large immovable equipment or facilities; information is intangible, and when transcribed onto paper or crystallized into electronic, magnetic or optical media, small and portable. To effectively secure such knowledge, a company must identify what the knowledge is, the components that comprise it, and how and where that information is stored.

Identification of a company's information assets is best done by the cross-functional IP Security Team, the executive team and legal counsel. The cross-functional IP Security Team can document the typology for each identified IP asset: engineering drawings, molecular combinations, formulation components, proprietary process documentation, source code and so on. Finally, an inventory needs to be conducted. Where does this information exist physically? Where does it exist digitally?

The inventory must be rigorous; copies of intellectual property that are unaccounted for undermine the entire safeguarding effort.

ASSESS RISKS AND PRIORITIES

Utilize a straightforward risk analysis to identify the intellectual property most at risk and prioritize the implementation of future controls. If the cross-functional IP Security Team is not experienced in developing a risk assessment protocol, bring in a non-team facilitator. Such a facilitator could be someone from within the company such as a finance or accounting individual, or an outside consultant with experience in developing and managing risk. The methodology and detailed capabilities of the risk analysis are less important than having a process that allows for the full exploration and consideration of risk factors and business timelines. Sample questions to pose include:

- Is the company in the process of signing a co-development partnership with another firm?
- Are there individuals who have helped create the IP who are in danger of leaving the firm?
- What will the cost be to the company's financials—short and long-term—if this drawing or formulation, etc., is made public?
- Is this for a product being developed or simply something to be licensed?
- Has this been patented or otherwise legally protected yet?

From the answers to questions like these, the company will know the level of risk its intellectual property faces and the priorities for controls to be identified and implemented.

IDENTIFY AND IMPLEMENT CONTROLS

There are multiple levels and complexities of controls and the IP Security Team needs to constantly be mindful of the three primary goals of intellectual property security: minimizing the risk of accidental disclosure, preventing industrial espionage and creating an environment that allows further IP refinement and innovation. All other goals are secondary.

Thus, when identifying potential controls to implement, ask:

- Which control is affordable over the long term?
- Which control will block internal knowledge

sharing?
- Who will monitor this control and its effectiveness?
- Which control can be automated?
- Which control can be verifiably reversed?
- Which control will allow for exemptions?
- How will this control be interpreted by a loyal staff member? A disgruntled employee? An external auditor or potential partner?
- Who will be able to quickly and easily identify this control and circumvent it?

There are no "right" answers to these questions.

In general, controls come in four forms: preventatives, alerts, alarms and triggers. Preventatives are items such as written policies, training, confidentiality and non-disclosure agreements, and management-based announcements or reminders; they also include default electronic security (for instance, requiring a username and password to log onto a computer prevents anyone not in possession of those two items from logging onto the computer).

Pros & Cons of an IP Control

Some companies utilize encryption methods to secure digitally stored intellectual property, whether on a portable computer (such as a laptop) or within the company's computer network.

Encryption deters the amateur from active theft and can make "criminal intent" easier to prove if the IP is ever stolen.

However, because decryption keys will need to be handed out to trusted individuals, the method is not foolproof. Nor is encryption of internally stored information cost-effective in the long run. Encryption adds a layer of overhead that must be consistently maintained, managed and taken into consideration. Simply overlooking this during a software or hardware upgrade can cause complete loss of the information, particularly if the disaster recovery routines were never fully tested and successful.

By seeking to understand the long-term consequences of a particular control, a company can avoid undue risks or burdens.

> ## Small Startups and IP Controls
>
> There are a number of controls that small companies can easily, and cost-effectively, put into place. Automated controls that come pre-installed in most software and computers include setting passwords on computers or electronic documents.
>
> Manual, people-based controls include escorting visitors, keeping non-monitored doors locked, training personnel, and implementing a "clean desk" policy (wherein personnel lock up their documents before leaving their offices and work areas).

Preventatives are designed purely as a deterrence. Because deterrence relies upon awareness for effectiveness, promote such preventative measures internally through training, memoranda and occasional reminders, formal or otherwise.

Alerts are the next level of controls and are designed to be proactive; alerts typically complement preventative measures. An alert is based on something that might happen (if condition X, then Y). These can be automated or manual. However, if not monitored, alerts simply become an after-the-fact piece of information. An automated alert familiar to many occurs when an incorrect password is entered three times in a row on a computer (condition X). In addition to the computer account being locked, preventing further log-in attempts, an alert is written to the computer log (then Y). An example of a similarly familiar alert, this time a manual one, is a phone call to a supervisor upon the arrival of a third-party auditor at a biotechnology company's front door.

Alarms occur based on something that is happening at that moment, such as the breakout of fire causing a fire alarm. For the computer environment, anti-virus software serves as a control to minimize the risk of virus infection or the installation of spyware; many anti-virus applications display an alarm to the computer user or the network administrator when a virus is either found or is in the process of attempting to install itself.

Triggers are reactive controls and typically are implemented based on threshold events—the automatic locking of a computer account across the network based on three or more bad passwords in a row is an automated triggered reaction; the lock-down of a building is a manually triggered reaction.

Because intellectual property is disclosed by people, solely relying upon automated controls to stop industrial espionage, minimize accidental disclosure or allow appropriate information sharing will not work. Controls that are manual or "soft" (people-based) must also be part of the equation. Escorting visitors through a building is a manual control that many organizations adopt.

For biotechnology firms that have at least fifty employees, even more effective controls are available, but such controls require a level of commitment that smaller biotechnology firms, particularly startups, will not be able to consistently deliver, thus rendering the control ineffective.

BALANCE ACCESS WITH ISOLATION

The third goal of intellectual property security—allowance for further refinement and innovation—is often the subject of much contention. Securing intellectual property is a balancing act, requiring significant subtlety and evolution over time. It is enough to state that innovation rarely occurs in isolation, while security is illusory in an open environment. Because intellectual property varies so widely, and the make-up of companies can lie between two partners in a garage and hundreds of individuals in virtual contact around the globe, there is no pre-defined industry "best practice" balance or a system that can define this balance.

There are, however, two methods that can provide an initial estimation of where the best balance lies for a biotechnology company: legal counsel's analysis of patents vis-à-vis possible discovery motion requests and a comparison of proposed IP access to the numbers of people typically involved in financial fraud.

PATENTS AND DISCOVERY MOTIONS

A company's legal counsel plays the defining role in drawing out a viable balance between access and isolation of intellectual property. By leveraging knowledge of required patent submission information, publicly available patent information and likely requested information (from discovery motions) in the event of product liability litigation due to personal injury, a biotechnology lawyer can chart a path identifying the various pieces of intellectual property comprising the product's entirety versus those IP elements that might be reasonably requested in litigation. In other words, by drawing a distinction between the IP as a whole and those components that, if publicly disclosed, would not necessarily lead to either the ability to replicate or reassemble the IP, legal counsel can sketch a broad set of potential boundaries within which to construct the access versus isolation balance.

This process can be illustrated by looking at a simple formulation (be it for a single active pharmaceutical ingredient, a set of complex molecules, etc.). A formulation has a number of characteristic specifications like ingredients, order of mixing, expected viscosity, appearance, temperature (or ambient temperature required for initial creation versus storage), and so on. In such a case, disclosure of the ingredients and even their specific amounts might be important in a product liability litigation but may not constitute all of the critical intellectual property elements if the order of combination, the time involved, the process and ambient environmental factors play a distinct role in the formulation's creation. While that information may eventually come out under judicial proceedings, for the purposes of striking a balance between what needs to be isolated and secured internally within the company versus what aspects of IP can be more accessible, this is the type of analysis that provides an initial framework. In the case of the formulation, this segregation of IP elements allows the biotechnology company to outsource some components of the formula creation (perhaps the first 4 ingredients that do not rely upon time, process and ambient factors), but still retain other components necessary for process completion. If the company has multiple projects occurring simultaneously, similar component sourcing between projects could occur amongst its various labs and outside partners.

FIVE-PERSON QUESTIONING

The second, more case-by-case approximation of the balance between access and isolation relies upon drawing inferences from the number of individuals involved in well-documented episodes of financial misconduct within companies, such as Enron, and from questions frequently asked by auditors such as, "Who is allowed to authorize changes to financial records?" Typically it requires five individuals (inside and/or outside the company), who are in regular contact with each other, to learn enough specific details of the intellectual property to be able to piece together the IP to allow solid replication or reengineering.

The best of both approaches would be to utilize the company's IP Security Team and legal counsel to lay out the initial components of the intellectual property, and then draw upon the second method, the five-person access rule-of-thumb, to assess potential activities such as selecting outsourced suppliers or partners.

APPOINT AN INVENTOR "GATEKEEPER"

Even if the controls implemented by the company are perfect, the balance struck between information access and isolation is tight, and legal counsel has been assiduously followed, executives need to be realistic concerning their ability to keep intellectual property secure. Given enough time, motivation and resources, an individual determined to obtain a company's information assets will invariably succeed.

Thus, biotechnology executives need to be open to less conventional means of retaining valuable information. Asking a scientist or engineer to serve as a Gatekeeper is a creative way to leverage the people at hand while also putting in place an IP theft (or disclosure) deterrent challenging to surmount.

The ideal Gatekeeper is a senior scientist or engineer—especially someone within five to eight years of retirement. These individuals typically have the best set of experiences, skills, maturity and earned positions of respect (formal or informal) to assess threats or opportunities presented by information sharing. Their goal is not to aggressively control information, but rather, like a

> ## Small Startups and Senior Gatekeepers
>
> The small biotechnology startup is unlikely to have senior scientists and engineers to draw upon to serve as Gatekeepers. Thus, such companies will need to look outside the immediate founders and/or staff to adopt the Gatekeeper strategy.
>
> Avoid the use of non-company accountable individuals (such as lab manager contractors, engineering design firms, outside consultants, etc.). The best Gatekeepers need to have some personal "skin in the game."
>
> If the startup is funded through capital ventures, request a Gatekeeper be provided by the investor; if the startup is not externally funded, explore the option of using an individual who currently serves on the company's scientific advisory board and is under contractual obligation.

gate, to channel it, allowing just the right amount into the mix to accelerate innovation and improve discovery.

Encourage the Gatekeeper to contribute to scientific literature, speak at conferences, serve on panels, explore the available literature voraciously and to translate those experiences and the information gained back to their colleagues as trends, opportunities, and risks (or even threats, in the case of a competitor). The Gatekeeper will be able to draw solid conclusions from the available public literature, the activities of colleagues, and, because of his/her awareness of this broader industry spectrum and the biotechnology company's internal affairs, serve the company's senior executives, governance board and general counsel. Of the many scientific ideas generated in a biotechnology company, the Gatekeeper is crucial to separating the viable from the indifferent.

Complement the Gatekeeper's work by creating a "knowledge team" for him/her to lead. An interesting and often fruitful team consists of sales, marketing, business development and customer relationship personnel who are tasked with capturing and continuously monitoring information about the marketplace, customers (current, former, and prospective), suppliers, pilot operations, focus groups, and competitors. The Gatekeeper can then draw upon this information to help organize clustered contexts in which intellectual property components can be effectively shared to foster innovation.

The Gatekeeper also acts as an early warning during development collaborations and/or preliminary negotiations. He or she recognizes the types of inquiries driven from a wish to help the partnership and co-development succeed versus those questions more derivative of a desire to capture all the details and nuances of a company's intellectual property.

ACCEPT THE TWO CHANGES OF PERSONNEL

Even during the early stages of a biotechnology firm, the reality is that personnel change is unavoidable; in fact, depending on the nature of the company, whether it is multi-product oriented or virtually built, entire groups may come and go. Supplement the controls and the Gatekeeper with proactive processes to tackle the two most common changes: the entrance of new personnel and the departure of experienced individuals.

ON-BOARDING

For new personnel joining a biotechnology company or product development effort, have an on-boarding process that blends "hit the ground running" aspects with a streamlined IP awareness and security review. Include all the elements that first went into developing the security program for intellectual property, from the IP primer to policies, preventative measures, the roles of the various individuals, and the teams involved. Have the IP Liaison and either a product development's project manager or the company's Gatekeeper run this portion of the on-boarding process. Larger biotechnology businesses will also want to incorporate other functional areas such as records management, IT, quality assurance and so forth. When individuals shift roles within a company, such as promotion to a new level of leadership and responsibility or a move to a department more actively involved in IP creation or management, consider implementing an abbreviated version of the on-boarding process involving the IP Liaison and the new supervisor.

Debriefing

From the standpoint of securing intellectual property, there are three steps to debriefing departing personnel:

1. The IT department needs to take an electronic capture of all product-related documents and files stored, created or otherwise modified by the individual in question. This archive should be retained in "read-only" fashion for one year after the departure.
2. Have the product development project manager and the IP Liaison meet with the individual and review his or her role in the project and discuss what was worked on, if there is any private feedback for the development effort; focus on knowledge about the product development activities, risks and/or opportunities that the individual may have that might not be captured somewhere (simply asking is often the most effective approach).
3. Ask about the possibilities the individual sees for the development effort such as other product opportunities that have yet to be explored.

The debriefing session is not a search for hidden clues or secret agendas; rather it is an attempt to capture information one-on-one that an individual may be in reticent sharing with a larger group or may otherwise not have considered significant to the biotechnology company's efforts.

Depending on the circumstance of the individual's departure, the company may also want personnel from Human Resources, Legal and/or Records Management to be involved. Regardless of the nature of the departure, have the IP Liaison conduct a brief review of the IT snapshot of the individual's files prior to the de-

> ### Startups and Debriefing
>
> Anecdotal evidence and informal experiences indicate that small startups suffer disproportionately from the loss of personnel, losing precious time as well as experience; rare is the small startup with a strong personnel knowledge transfer infrastructure.
>
> For biotechnology startups that feature significant specialization, personnel departure impacts can be deeply felt and carry added risks; the need for effectively securing IP during departures is in these cases is acute.
>
> There are several cost-effective tactics startups can explore:
>
> - Utilize a local human resource service provider
> - Purchase and adopt a records review template from the Association of Records Managers and Administrators International (ARMA)
> - Structure retainer-type contracts or severance packages to allow the company breathing room
>
> Companies targeting ISO compliance can even draw upon elements found in ISO-17799 and ISO-15489.

briefing. The IP Liaison is in the best position to quickly identify items that warrant additional questioning during the debriefing.

CRAFT A DEVELOPMENT AGREEMENT

At a minimum, the IP Liaison, if not the full cross-functional IP Security Team (chaired by the company's legal counsel), needs to be deeply involved in creating protocols for information and knowledge management flow between any collaborating organizations, as well as guidelines for information distribution back into each organization beyond the project teams and sponsors.

If the contract between the organizations does not specify this detail, craft a Joint Development Agreement (JDA) focusing solely on information and knowledge flow. This should be written within the constraints of the legal contract formalizing the collaboration between the two organizations. While the partnership contract always supersedes the JDA, frequently roles and processes from the JDA become contract amendments.

Key components and processes to address in the JDA are: communication, issue resolution, project reviews, roles, checkpoints, or stage-gates, and other types of information exchange and decision-making arenas that should be aligned between the collaborating organizations. If the contract between the organization does not specify intellectual property ownership aspects, utilize the JDA to spell these out, differentiating between initial IP that each party brings to the collaboration and ownership of new IP that results from the partnership.

If possible, consider creating an IP fact sheet as an addendum to the JDA that walks through the basics of IP ownership and rights—in essence, a simplified IP primer that also includes a disclaimer referring to the specific contractual agreements.

> **Startups and Development Agreements**
>
> Compared to deeper-pocketed established biotechnology firms and pharmaceutical companies, startups are at a distinct disadvantage when it comes to negotiating and securing an IP protective co-development alliance.
>
> There are two tactics that may come in handy for the startup biotechnology firm:
>
> - Utilize a local 3rd party arbitrator to help hammer out an agreement
> - Purchase a core template from a generic service provider over the internet
>
> There are two template types quickly found: university-based development agreements published on the internet and software development agreement templates. Many of the IP protection and ownership mechanisms in software development agreements are similar to, and thus easily transferrable to, biotechnology.

While the JDA is not a legally binding document, it is a set of operating principles and agreed upon guidelines between the members of the two organization's teams, and as such should form part of the introduction to the collaboration for any new personnel, employee or contractor, as well as provide guidance for personnel either changing roles within a company or departing the company altogether.

SUMMARIZE IP TWICE

To further enhance innovation while safeguarding intellectual property, create an intellectual property summary document in two distinct parts so that it can be shared with two diverse audiences.

The IP summary's two sections are a high-level overview and a more detailed, technical review typically written by both the IP Liaison and the Gatekeeper. For individuals who are not scientists or engineers, there is no need to provide the detailed, technical review.

The high-level overview should be approximately 200-300 words within 12-15 sentences, targeted to an executive or non-technical reader. The IP summary should be similar in tone and verbiage to the IP primer. The overview includes: a summary of the company's goal and intended product, the differentiation of this product in development versus products already on the market, the current stage of development, and any specific IP previously or otherwise publicly disclosed (such as patents, trademarks, etc.).

The detailed, more technical section is significantly longer and is specifically written for the product engineer and/or scientist who will be actively working with the intellectual property. Include drawings, photographs, formulation descriptions, process workflow diagrams, proof of concept or demo versions, and so forth as appropriate. For collaboration-based IP summaries, identify key contractual terms and conditions, provide a current copy of the JDA, and point out where to go and whom to see with questions. Clarity must be balanced with brevity; the goal is to

> **Startups and IP Summaries**
>
> There are two additional reasons for the startup biotechnology company to draft IP summaries:
>
> - IP summaries can aid the quest for investment capital
> - IP summaries—depending on how they are written—can serve the dual purpose of categorizing a firm's IP (as part of developing a secure information framework)
>
> For a sample format, the intellectual property marketplace, Yet2.com (at the time of writing) provides a downloadable "Tech-Pak Form" that is easily adopted by the startup for internal IP summary use.

allow the new personnel to rapidly and effectively integrate and contribute to the IP of the company (or companies in the case of a co-development alliance).

IP summaries provide two subtle safeguards. Curiosity about the company's new product ideas or innovations is usually satisfied with a top-level summary; this prevents idle curiosity from spawning more intentional "fishing expeditions" that risk either greater IP exposure and/or the triggering of security responses not necessarily commiserate with simple curiosity. Secondly, IP summaries provide enough detail to scientists and engineers focused on leveraging the information for further innovations; this will inhibit most desires to dig more deeply and thus risk the unwitting disclosure of confidential details of the intellectual property in the context of a new discovery or invention.

CIRCUMVENT YOUR OWN LIMITATIONS

Biotechnology executives need to recognize their own limitations when securing intellectual property; this is how they can find the means to circumvent resource restrictions, surpass expectations, and assure the company and its investors of long-term IP dividends.

There are four types of IP security limitations inherent in every biotechnology company:

 a. Resources
 b. Expertise
 c. Stratagems
 d. Techniques

Resources encompass the constraints of funding, personnel numbers and types, time, equipment and technology, and advanced knowledge of future marketplace changes and collaborations. Expertise includes awareness of industry trends and best practices, hidden IP security pitfalls and the subtle nuances available only through experience. Stratagems comprise knowledge of potential options, innovative strategies and flexible field tactics to maximize a company's strengths and compensate for its weaknesses such as lack of funding or automated security capabilities. Techniques encompass specific tools, tricks and process components that can be drawn upon to target the needs and desires at hand within the available constraints.

Biotechnology executives who understand these limitations will be able to circumvent them and effectively secure their company's intellectual property.

Methods of addressing these areas of weakness include bringing in outside consultants, adding staff with expertise that compensates for some of these limitations, or looking to outside resources on an as-needed basis to provide help.

Adding personnel with expertise in these areas may require more resources to manage appropriately, requiring more efforts to be diverted from long-term product development activities. There are no advanced educational degrees in intellectual property security to provide recognized accreditation, nor are there assurances that the individual's previous experiences or capabilities will translate effectively into the biotechnology company's interests.

Looking to outside resources such as independent consultants, external law firms or security vendors is the more commonly adopted tactic, at least initially. However, it is only less risky and less costly if actively managed, and avoids long-term staff augmentation and other forms of entanglements. Critical to such an intention is implementing a "knowledge transfer" component to any contract with outside resources; even if the company decides not to develop its own skills to secure its IP, the biotechnology firm must have enough knowledge to grant itself leeway in transferring business between outside professional services without undue risk to its intellectual property.

There are two parts to a cost-effective, knowledge-transfer based strategy utilizing external expertise. The first requires company executives to attend (or schedule in-house) a workshop covering IP security basics and industry best practices, allowing for personal questions to be answered. Out of that workshop, the executives can then identify the cross-functional IP Security Team members. Depending upon the company's comfort with information management security and risk assessments, an outside expert can be brought in to serve as facilitator for the risk assessment and prioritization process. Small startups may want to request their investors provide help in this context, either through an outside consultant or through an experienced member of the venture capital firm itself.

The second component of external expertise knowledge transfer can occur over the course of implementing the remaining strategies identified in this chapter. Look for an outside expert to place on retainer, drawing the individual into the company only on an as-needed basis. Second, hire a third-party auditor (not the retainer-based independent consultant) to assess the controls in place. Either the retainer-based consultant or the auditor should also be able to provide the biotechnology firm a high-level strategy plan for long-term sustainment and improvement of IP security as the company evolves. Ideally, such a plan will incorporate potential trigger points, such as a future IP licensing arrangement or co-development alliance, wherein the company may want to either bring in the trusted expert advisor for additional insights and improvement suggestions appropriate to the business development at hand such as prior to commencement of a deal or as part of the joint contract or development agreement. Be cautious about utilizing the same outside resource as both advisor, auditor or counselor; segregation of duties to minimize self-interest and temptation is very difficult in such situations. Biotechnology business development demands balanced advice for companies, shareholders and patients—not to mention the science itself—to advance.

FINAL THOUGHTS

Intellectual property is the essence of the biotechnology business.

Securing intellectual property represents the greatest concern for biotechnology businesses. For biotechnology executives, it is the way to market release and bottom line growth; for biotechnology investors, it is the difference between a successful initial public offering and an ignominious bankruptcy notification; and for future biotechnology patients, it is the basis of life and death.

Aggressive Intellectual Property Strategies
Gene Rzucidlo and Stefan Miller

Gene Rzucidlo has extensive experience in the preparation, filing, prosecution and appeals of patents for a diverse array of technologies. He is well-versed in patent procurement as it relates to both domestic and international patent law through his extensive trial experience involving patent law and patent office procedure. Gene's experience provides keen insight into patent interference strategies, re-examinations and reissues and he represents clients in patent prosecution, patent validity, infringement, clearance studies and opinions. His clients include established and start-up companies. Gene can be contacted at *gcr@hunton.com*.

Stefan Miller, Ph.D., is a registered patent agent whose work focuses on drafting and prosecuting patent applications in the U.S. Patent and Trademark Office and in performing legal research for litigation. Stefan can be contacted at *smiller2@hunton.com*.

As demand increases for biotechnology, the push for new technologies and innovation requires the prudent biotech business to invest in offensive and defensive Intellectual Property (IP) strategies. Biotech's potential brings competitors into an increasingly crowded landscape; therefore, an aggressive IP strategy can provide a means towards claiming a foothold as well as expanding business territory. This competitive landscape offers unique challenges as a knowledge-based marketplace. A company must actively seek innovations to build a portfolio of strong IP protection, while extracting maximal value from its IP.

This chapter will provide practical tips for aggressively using IP both offensively and defensively. An aggressive IP strategy envisions IP as more than a means to protecting one's intellectual assets from appropriation but also as a means towards profits. Such a strategy develops business practices which limit or obviate costly litigation or the inadvertent loss of IP rights. This chapter will also discuss the advantages and disadvantages of various strategies for a company in deciding when to litigate or when to license, or when to actively seek to stop competitors from issuing patents.

As with any advice, certain caveats are in order. A company should develop diverse strategies structured for the various business objectives of its businesses and actively prioritize their goals. IP is not a monolithic idea that requires a single strategy for the entire company but should be tailored to the specific needs and goals of its divisions. For example, one particular business unit may be interested in obtaining patent protection in order to license with competitors in the field, whereas another unit may be more keen on utilizing its patent leverage to sell products. There is much more intellectual property protection available than is often taken advantage of by biotech companies. Development of better business practices are one underutilized area of IP protection that is often overlooked amongst the trees in the forest. Further, IP protection should also be pursued with a purpose in mind. The aggressive use of IP envisions each IP asset as a means to an end. As such, success should be measured alongside these specific business objectives. There are three critical components in instituting an IP strategy that supports a company's business objectives.

First, a company must actively utilize the statutory and common law protections afforded to intellectual property. In many instances, this requires that a company engage in identifying innovation and using best-practices to ensure IP protection can be secured. This defensive use of IP involves seeking IP protection before others. Second, a company should aggressively seek areas where it can claim pioneering status.[1] This chapter will discuss the use of strategic inventing to this end. Third, a company should extract all value from its IP, which includes the offensive use of IP in-licensing, cross-licensing, or litigation. In many situations, IP can also be used both offensively and defensively such as maintaining a strong and actively growing patent portfolio to be used against other companies attempting to assert patent rights.

In summary, a host of interrelated strategies, business practices,

---

1   There are many possible definitions for when IP protection covers a pioneering innovation, but for practical purposes, an innovation is pioneering when competing innovation is lacking.

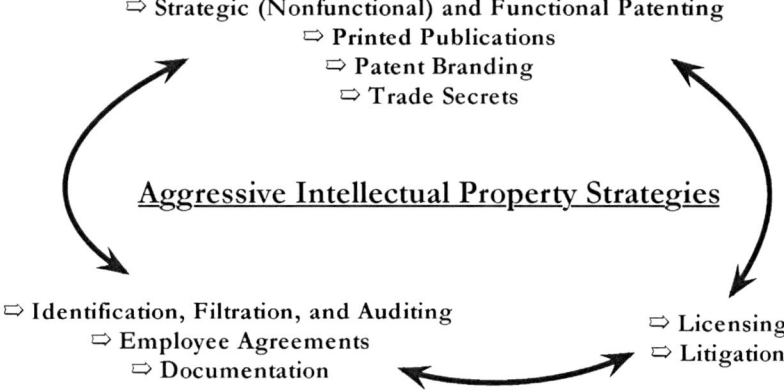

Figure 1: Aggressive intellectual property strategies

and legal concerns interact to provide a network of a means to an end. Figure 1 provides a simplified picture of the strategic concerns a business should incorporate. IP strategy facets are self-reinforcing. Therefore, companies incorporating multiple practices into their business plans will generate larger yields.

BEST BUSINESS PRACTICES

IP assets have a life-cycle. They are first born as ideas, which alone are unprotectable. Next, they develop into specific innovations that are protected legally by a quadfecta of statutory and common law systems: copyright, trademark, patent, and trade secrets. IP can be a source of profit, either by selling the product or by enforcing IP rights through licensing or litigation; however, a company should first ensure IP asset development and growth is nurtured and sustained by best business practices. Best practices require an understanding of employee agreements, proper documentation of innovation, and use of trade secrets. They also include means for the identification of IP assets, the filtration of low-value ideas, and for self-assessment. Further, the use of strategic publications and patenting, coupled with patent branding, provide the framework for successful licensing/litigation schemes. These practices are discussed next.

Employee Management

A company's employees are the production arm of a company's IP. As an initial matter, an employer should ensure proper management of its legal relationship to its employees. A company can define and enforce an employee's legal obligations and responsibilities using employee contracts to allocate ownership of IP. Patents invest in owners and copyrights invest in authors. Affirmative steps must be taken to obtain exclusive rights to these properties before they are developed. Typically, employee contracts and work-for-hire agreements are used. These arrangements explicitly state employees are hired to invent, and all inventions or other intellectually property assets belong to the company. The company should also include statements outlining an inventor's obligation to assign inventions to the company.

Employees should be informed of a company's confidentiality policies as well. In addition to ensuring confidentiality is maintained by in-house practices designating intellectual property asset disclosure as privileged and confidential, a company must ensure these policies are understood and enforced. As part of an employee contract, new hires should also sign a statement identifying the company's confidentiality policies. These policies should be periodically redistributed to all employees in written form. This is particularly important when a company utilizes trade secrets.

Trade Secrets

Trade secrets include proprietary information a company decides to keep secret because of its economic value to the company. Trade secrets maintained in secrecy provide nearly indefinite protection, but they are particularly vulnerable in the modern marketplace. They offer no protection against the use of information a rival company has independently created or even information that a rival company has accrued through reverse engineering. This form of IP is difficult to maintain because employee/inventor turnover is often high in many industries. Even with explicit agreements not to divulge propriety information, many employees and inventors nonetheless may divulge information on a subconscious level. Even while in-house, employees may divulge information inadvertently in publications or presentations, especially when trade secrets are not specifically identified as proprietary. Lastly, while litigation is possible when corporate espionage has occurred, without a strong initial foundation, proving trade secrets were actively maintained/protected can be difficult.

DOCUMENTATION

Ideas alone are never protectable. Copyright protects the expressive elements of a work whereas patents protect the innovation as reduced to practice. As an idea grows from an intangible conception to a final product, a company needs to ensure researchers and inventors provide detailed logs and documentation of all efforts.

DOCUMENTATION IN A BIOTECH LAB

A researcher in a biotech lab needs to record procedures and results in a bound lab notebook. Daily logs alone, however, will not provide sufficient documentation in a later infringement action. Corroboration is key. All documentation needs to be signed by a party who is not affiliated with the research and who can understand the contents. Between entries, large, empty spaces in a notebook should be crossed out, dated, and initialed. Errors should not be erased (indeed pencil should never be used) but crossed out with single lines. These technical best-practices should be rigorously enforced.

A company should even document its efforts towards enforcing these best practices and provide employees with manuals describing best practices for intellectual property. Invention disclosure forms should be provided to inventors even before a potential IP asset is identified. These forms will present a standardized means for inventors to document all aspects of the inventorship process such as the what, where, when, why, who, and how. Even if a company does not routinely seek IP protection for its inventions, clear and consistent documentation practices are a low-cost means for minimizing the risk of losing critical IP rights and aid in defending against possible litigation over such rights.

IDENTIFICATION, FILTRATION, AND SELF-ASSESSMENT

A failure to educate employees the rudiments of intellectual property (i.e., what it is, how it should be protected, and how to avoid infringement), can cost a company. Frequently, mistaken beliefs as to what constitutes intellectual property result in inadvertent disclosure of trade secrets or the publication of information which could act as prior art against a later attempt at patenting. While business best practices can prevent unwanted disclosures, protect against infringement, and even help a potential inventor recognize what can and cannot be protected, a company should ensure all personnel involved in the creation or management of IP are fully conversant in the critical components of IP rights. Integrating short seminars or retreats into a company's human resource program can provide a simple means of ensuring employees are competent in the rudiments of IP. Inasmuch as lawyers are frequently required to participate in continuing legal education (CLE) classes, such in-house education can also improve employees' ability to recognize IP and to convey it efficiently. In conjunction with such a program, an idea clearing house can ensure innovative ideas and recognized IP opportunities are channeled to the persons and divisions in which their potential can be most fully realized.

IP asset management is also crucial for a growing business. A company should aggressively re-evaluate its position on a periodic and frequent basis. For example, a company should survey its business units to determine which IP assets are considered most important and how they would allocate funding amongst these key technologies. If a business has international concentrates, the survey should also ask which countries the particular unit deems most important as far as IP protection. These surveys can provide a foundation for proper patent management. Patent management should also seek to identify low and high value patents. High value patents typically constitute core technologies that preserve a company's position in the marketplace. Other high value patents might also include patents which prevent competitor entry into the marketplace. Economic considerations have long suggested that products entering the marketplace without a competitive advantage will lose profit over time. Patent protection should be evaluated according to the strength of competitive advantage it affords. This competitive advantage need not only be a functional advantage in the company's product, but can also be derived from the strategic patenting and marketing of non-functional differences.

EVALUATING THE STRENGTH OF A PATENT PORTFOLIO

There are several benchmarks a company can use in evaluating the strength of its patent portfolio, but, typically, a patent portfolio is strong if it contains a series of patents having patent claims covering different aspects of the business's core technologies. Such a portfolio therefore has "blocking patents" which are hard to design around and thwarts competitor entry. The prudent use of "picket-fence" patents is also well-advised. These fence patents claim minor improvements likely to arise and make it difficult for competitors to design around the patent or institute blocking patents with minor improvements.

Regular IP audits provide the framework by which a company can plan the most effective use of their resources. Teams composed of individuals familiar with the company's current and future planned technologies should initiate audits. Audits need not and should not be limited to self-evaluations. The auditing period is the perfect time to evaluate the IP position of competitors because the company's own position is in temporal juxtaposition. IP audits are critical for start-up companies, because they ensure core technologies are adequately protected and appropriate fencing patents are sought. Established companies can use the IP audit process as a time to refocus their patent portfolio.

Further, newly acquired or created intellectual property should be evaluated to determine whether commercialization, production, or another use would infringe on the IP rights of others. These freedom-to-operate (FTO) investigations are crucial in preventing costly infringement litigation. Combined with regular IP audits, evaluative investigations also provide information about competition. Potential infringers of a company's IP rights are more easily identified after an FTO/IP audit, and potential licensees are discoverable.

THE LIFE CYCLE OF INTELLECTUAL PROPERTY

When an idea has been developed into a tangible asset, a company must be prepared to determine a course of action for protecting this asset. Trade secrets offer the potential for indefinite protection but with certain attendant risks. Alternatively, a company can seek patent protection affording a monopoly of limited duration.[2] With an aggressive IP strategy and the ability to produce definite and regular improvements over a company's prior innovations, additional patents with non-obvious improvements can extend the time a company can preclude others from competing. Ideally, a company will strive to develop a diverse and constantly expanding patent portfolio. Unfortunately, patents can be expensive and the prospect of periodically incurring the cost of protecting every minor, non-obvious improvement can lead a company to think that it can do no more than ride out the limited patent monopoly it has been granted. This is where an aggressive IP strategy can come to the rescue. Through the use of aggressive printed publications, a company can improve its chances of preventing others from patenting innovations it currently practices in one form or another.

PRINTED PUBLICATIONS

Patents are granted to novel and inventive innovations. Inventiveness is referred to in legal circles as a question of nonobviousness. In other words, even if an innovation has never been duplicated in its exact form by another, the inventor can be pre-

---

[2] *Copyright is sometimes an alternative, especially for companies in the computer industry where the utility of a software program is often so short-lived and the patenting process long in comparison. Software companies often use copyrights in addition to the gratuitous use of printed publications to provide IP protection.*

cluded from receiving a patent if the innovation is not sufficiently inventive from the viewpoint of prior art. The prior art is the mass of information that is in tangible form and publicly available. Good business practice would then advise an inventor to not create prior art which can be held against the company! For example, suppose an inventor discovers a new way of manufacturing a widget. In this hypothetical, the inventor has toiled long and hard to make this improvement and has created an invention which is so nonobvious as to easily "fly" through the Patent Office. In their excitement, the inventor posts a detailed description of the widget on the company website and even submits an article to a local Widget journal which promptly publishes it. The inventor returns to the lab and spends the next two years finalizing the invention before filing for a patent. Much to their dismay, an Examiner will reject the application on the grounds that the invention was publicly disclosed more than a year before filing.[3]

Such a situation might lead a company to consider instituting an outright ban on employee publications altogether. An aggressive IP strategy, however, can utilize this patent protection intricacy to its advantage. Consider another example, wherein an inventor has once again invented an incredible innovative widget manufacturing method. The inventor's company wisely seeks patent protection and this time the inventor does not reveal the invention in a publication. The company begins using the new method and soon the inventor discovers a few nonobvious improvements to the method which will save the company a small amount of money. The company wishes to use the new method but cannot maintain it as a trade secret because it is a method that is easily reverse-engineered. The company could once again go through the patent process but the cost might not be worth it.

---

3   *The Examiner could rely on the website posting or the journal publication. Websites are sometimes difficult issues because they might not be considered prior art as of a certain date without a date indicating when it was posted. This of course does not mean that web postings are okay as long as a posting date is not included! As discussed below, prior art such as this can be dated in other ways.*

An aggressive approach to this situation is to turn the inventor's earlier mistake into a must-do. The company's main concern here is that another company might patent the process and prevent the inventing company from practicing the minor improvement. As such, the company can use its IP knowledge to prevent others from patenting by disclosing, through publication, their improved process. This lays a prior art foundation for challenging a company seeking to patent the inventor's process.

Because publication can be a benefit on one hand and a detriment on the other, a company should institute an in-house clearing center for the dissemination of information to outside sources. Documents remaining in-house are not treated as publications when the routine business practice is to maintain these publications in secret. However, as suggested above, it is best to show the practice as explicit through the use of confidentiality agreements and disclaimers like those commonly found in emails.

But when is a document published? As demonstrated, the medium is not very important; rather, access is critical. A document is considered published when it is accessible to those interested and ordinarily skilled in the subject matter or art, who, with reasonable diligence, can locate it. Courts have held that even obscure documents, in foreign languages, resting in the back of a dusty bookshelf were sufficiently accessible.

Publishing

A company wishing to publish has a number of options. A website disclosure is quick, easy, and cost-effective. However, the company exposes itself as the source of information and care must be taken that all disclosures have clear and documented posting dates. The patent office offers a program called *Statutory-Invention-Registration* whereby a company can submit a disclosure that the patent office will publish in exchange for giving up rights to later prosecute the application to a patent. A disclosure in an SIR can affect related patent applications and there is a fee for publishing. Further, the SIR is effective as prior art as of its filing date and not its publishing date. A third option is to submit the disclosure to an outside publishing company. Cost is typically the only drawback to this option. Lastly, an aggressive IP company could submit prior art to the PTO examining a rival company's application. This should also be done with care. The company may not submit a description of why the disclosure is relevant and the Examiner may dismiss the disclosure after only a cursory review. Once examined, the Examiner's dismissal of the disclosure is accorded deference and as such may actually shore up the strength of a rival's patent should it be issued. Further, the company may amend its claims to work around the prior-art submission or simply argue that it is not pertinent prior art.

It should be noted that disclosure does not always obviate patent rights. In first case given above, the inventor's company waited two years before filing a patent application. The inventor would not have been barred by the PTO had he or she filed in less than a year. Indeed, the later improvement and disclosure could have been turned into a continuation in part of the first invention as long as the disclosure was not more than a year before filing. Also note, others may still be able to antedate the inventor's publication date by showing earlier invention. This is not always true in foreign jurisdictions. Indeed in some jurisdictions, the filing date is the date against which prior art and priority of invention is measured. As such, publication may be a bar against patenting even if publishing was done a day earlier. However, if an inventor doubts he or she will enter the market in such a jurisdiction, a defensive patent publication will ensure no one else receives a patent either. To prevent accidental loss of rights, an in-house publication clearing house can be used to review and approve all outgoing publications.

STRATEGIC PATENTING

Patenting innovation in the biotech field is limited by the scope of prior art. If the company's innovation is the same as something disclosed in a journal publication or patent, then it is clear that exclusive rights will not be granted even if the innovation was independently conceived. The prior art can also be used to prevent patenting of an innovation that is not exactly disclosed in any one publication, particularly when the improvement is minor and the examiner can find publications which suggest the improvement. Some biotech advances may be minor incremental improvements over the prior art. A major improvement is more likely to provide strategic positioning opportunities for IP protection because it will be difficult to design around and thereby yield an increased scope of competitive advantage. Further, companies typically seek IP protection for minor advances after their development. While it makes business sense to evaluate potential innovations when presented, such innovations have been developed according to their fit with a company's current business objectives and not with respect to their patentability, i.e. their ability to "overcome" the prior art.

A focused approach on inventing to develop a strong IP portfolio turns the traditional model on its head and is a useful adjunct to the traditional model. Instead of evaluating discoveries after they have been invented, research is directed into areas where strong IP protection is certain because of, for example, a dearth of prior art. This "white space" in the prior art presents opportunities to dominate a business territory. Indeed, this "white space" is an invitation to a company's competitors. Strategic patenting, however, involves more than seeking out areas where a company can obtain market dominance because of a strong patent position; it is also a means of denying the competition such opportunities. In this regard, a company may seek patents it has no intention to implement because these patents prevent competitor entry.

How does a business go about defining a strong IP position? A business needs to determine how to make its product sufficiently different so that patent protection is broad. The product does not have to be better. The PTO will not deny a patent because it does not provide an advantage over the prior art but only if it is sufficiently different.

An Example of Strategic Patenting

Pharmaceutical companies have utilized this to their advantage in extending the IP protection for their products. Many compounds exist as enantiomers; simply put, a compound is like a hand in that can be left-handed or right-handed. These left- and right-handed compounds are patented as a mixture. The active form (either the left-handed version or the right-handed version) can be separated and patented as the pure form. In combination with a marketing strategy, consumers will buy the "new" non-generic. Non-functional differences can thereby provide a strong patent stance. Consumers will often attribute non-functional differences with improvements in quality, which thereby increase sales and demand.

How does a company put strategic inventing into practice? First, a company should ensure that its researchers are well-versed in the prior art of their field and related fields. With this in mind, technologies can be evaluated and prioritized according to how well-known or used a technology is in the company's field. A technology having potential to be adapted for use in the company's field but not adapted in any of the prior art of the company's field is a potential vehicle for developing broad patent protection. Look for prior art in the field which suggests a method or process doesn't work or won't be suited for a particular purpose. These references are evaluated by the PTO as strong evidence that an improvement is non-obvious. Obviously, the company's product will have to be functional to sell, but strategic patenting looks for aspects of non-functional difference to develop patent protection. In other words, a minor innovation can be strengthened in regards to its patent scope by including non-functional differences which do not impede the underlying functionality. These non-functional differences can also be used as entry-points into a saturated market by marketing, which highlights these non-functional differences as indicators of source or quality. This is a critical point. The non-functional differences must also be ones which consumers and target businesses recognize. Nonetheless, patents that do not claim dominance over a particular field are still powerful adjuncts to a strong patent portfolio, particularly if they are used with an aggressive licensing/transfer scheme. Licensing is discussed below.

Patent Branding

A larger, well-known company can generally rely on the industry to independently reach the conclusion that its IP portfolio is strong and has considerable value. The smaller company must rely on a more aggressive approach to convey its strength. Marketing helps establish small company as IP powerhouses. Some larger companies have developed recognition for their IP assets by sheer quantity and not necessarily quality. As such, the smaller company should first focus on keystone patents that provide broad protection and strength. Valuation of a patent portfolio is controversial because no clear metrics have emerged, but a company can demonstrate strength by having defended its primary IP assets in litigation or by demonstrating a broad licensing network. In addition, with growth, a company can acquire a sufficient number of patents such that the individual quality of any one patent is usually not questioned because the mere quantity of protection dissuades potential infringers. Lastly, a company's patent assets should not be considered in a vacuum.

Developing good-will and associating this good-will with a company trademark can extend goodwill to a company's IP assets. Potential licensees entering dealings with a well-known mark demonstrate a subconscious respect for the company's assets. For example, GE has one of the strongest patent brands on the market. By sheer quantity, it need not market its IP portfolio directly, but it still engages in marketing directed to the public at large with such advertising slogans as "We bring good ideas to life." At its simplest, patent branding instills a message to competitors and licensees before litigation or before negotiation that the company values IP assets highly. Patent branding therefore adds value to IP.

Provisional/Nonprovisional Applications

An identified IP asset must be protected either by maintenance as a trade secret or by filing a patent application. In the US, there are two potential vehicles for accomplishing the latter: a provisional or non-provisional application. A provisional application reserves patent rights for a period of one year at which time a non-provisional application must be filed or patent protection is lost. A company may use provisional filings as an effective strategy in determining the value and viability of a potential innovation. It also demonstrates to potential investors that an innovation is on track for patent protection and can provide a period wherein a company can accumulate capital without having invested too many sunk-costs. At the end of the provisional period, a company may choose to abandon the application if market or other factors indicate the patent protection is not worthwhile. It is important to remember, however, that a provisional application is not merely a placeholder. While it is somewhat less complex then a non-provisional application, it is still a legal document. For example, while claims are not required, the document must fully describe and enable the invention to be patented. Broad claims can be added nonetheless and should be. It is easier to narrow the scope of an application when filing the ensuing non-provisional application than the reverse, since aspects which are not described in the application cannot receive the priority of the earlier filing. Incorporating claims ensures the broad scope of your patent is clear to the company and to the patent office.

Lastly, in some situations a company may have identified an asset for which patent protection is needed quickly in order to maximize value. This is especially true in technological areas (including biotech), which advance quickly rendering older technology moot in a matter of years. In such situations, a company should consider filing a nonprovisional patent immediately. Patent prosecution can take three years or longer and a nonprovisional application begins the examination process. In some instances, such as when the application is related to environmental, energy, or counter-terrorism concerns, an applicant may file a petition to expedite the process. Further, drafting a patent application with an experienced legal professional ensures the specification is clear and concise yet descriptive of the full scope of the innovation. Proper drafting can minimize the number of back-and-forths between the patent examiner and the applicant.

LICENSING

A company must be able to successfully leverage its IP assets if it wishes to garner maximum profit. A company should utilize these assets to either exclude competitors from the company's product's core market or to demand compensation for the use of the company's technology or know-how. With regard to the latter strategy, patent licensing is one means to this end. This section will discuss the benefits and disadvantages of licensing and some of the how-to behind it.

A company should first analyze the activities and IP protection of its competitors. This can be done by instituting regular IP audits of the company and its competitors. By evaluating the strength of the patent portfolio, concerned parties can determine whether it is broad enough to launch an effective licensing program. For example, management should be able to determine whether the scope of the patent portfolio covers all aspects of the product. Indeed, if the company's patent portfolio does not cover existing products, potential licensees may instead seek to simply copy the company's product instead.

Licensors should also be mindful of the risks inherent in initiating a licensing program, as negotiation with potential licensees may open a company up to litigation. A licensor should also recognize that not all licensees are created equal. A company manufactures the company's product but with inferior quality can generate negative good-will that might transfer over to the entire product line. Nonetheless, some licensees might be better positioned to exploit the innovation, and a new company can use incoming royalties and other licensing revenues to grow their business. Careful negotiation and proper licensing arrangements are therefore essential.

A company wishing to license its technology has numerous options in arranging a licensing program. Before licensing can begin, however, a company should identify the subject matter to be brought to the negotiation table. Does the company plan to license only the claims in the patented assets or will it also provide know-how or information protected as a trade secret? Further, the company should determine provisions in case the patent is later deemed invalid or unenforceable. Licensees will typically seek escape clauses that are broad in scope and licensors will aim for none. Generally, in-between provisions can be negotiated which provide indemnification against invalidating art which arises prior to the date of license.

A company should also adequately determine the value of its IP assets. Overcharging for a license can result in no licensees, and, with sufficiently valuable assets, may induce litigation leading to patent invalidation. Of course, undervaluation results in lost profits. Further, potential licensees should be screened to determine whether they will be able to fully develop and market the company's goods. A company can use incentives such as milestone payments like credits against royalties due when a licensee reaches certain deadlines. Attempting to contract hard-working behavior can result in disappointment, since diligence clauses can be vague or hard to enforce. Lastly, a company should be careful how it approaches potential licensees. Improperly drafted letters of interest, which can be interpreted as indicating a 'license or you infringe' approach, can allow a licensee to file a declaratory judgment action. With these preliminaries, a company can evaluate the type and scope of license it will seek to negotiate.

LICENSE TYPES: AN OVERVIEW

There are five main license types in the IP field. A *non-exclusive license* grants a licensee the ability to practice the company's technology but not the right to exclude others from practicing it. From the licensee's perspective, this is not a favored arrangement. Increased competition lowers the profit margin a licensee can expect. However, if the company is licensing a core-technology, a licensee may enter into a non-exclusive licensing arrangement. An *exclusive license* is the opposite and is particularly risky for the licensor. By its terms, the exclusive license prevents the licensor from practicing the invention as well. This imposes the risk that the licensee will not properly capitalize on the technology's potential. One in-between arrangement minimizing risks to either side is the *sole-use license*, which grants an exclusive license to the licensee but the licensor maintains the right to practice the invention itself. An alternative arrangement is the *field-of-use arrangement*. Here, a nonexclusive license is granted to practice the technology but the scope is restricted to a particular field or application. Lastly, a *know-how license* may also be arranged. In this arrangement, the licensor also grants its technical know-how as well as to the ability to make or use the technology.

The financial consideration for a license is also a critical aspect in negotiating a licensing agreement. Most companies include provisions for up-front technology fees which act as an advance upon future royalties that are not associated with or tied to actual performance. Alternatively, shares of stock in the licensee's company can be transferred as equity. Smaller companies can thereby retain cash, and the equity gained can appreciate. A company should carefully consider the potential implications of such an arrangement however. Equity transfers may include the right to a seat on the licensee's board of directors. While this affords the licensor a measure of control over the licensee's company, it might also create a conflict of interest. Another alternative is a fixed price or single payment license. This type of license is independent of performance. However, a company should be careful in such arrangements, since failure to retain rights in the technology, through an explicit agreement, can be construed as a legal sale or assignment of one's technology. As such, the license should explicitly disclaim these arrangements. Lastly, a company can institute a royalty sharing scheme, but should be sure to institute a means, such as an audit, to determine all payments due are indeed being received.

INFRINGEMENT ACTIONS

A company utilizing aggressive in-house strategies lays the framework for enforcing its patent portfolio effectively against competitors. During the crucial IP audits discussed above, a company must scour the marketplace to identify infringing competitors. Patent rights create a right to exclude but it is a self-policing right. A company should also keep in mind the likely cost of litigation which is usually high and is a drain throughout the lifecycle of the litigation, which can last several years. The advantages of a successful infringement action are two-fold, however. The successful claimant can win monetary damages and an injunction which shuts down the competing business entirely.

Litigation can also be used to preempt litigation threatened by a competitor. If a competitor aggressively seeks a licensing deal or intimates a patent in the innovating company's portfolio is invalid, the company can seek a declaratory judgment against the competitor for validity. This provides the advantage that the company can choose the forum in which to sue, which may result in a settlement with favorable licensing terms. If the marketplace consists of smaller and larger infringers, it is wise to seek litigation against smaller companies first since they are more likely to settle before costly litigation by arranging a licensing scheme. Successful litigation builds on the framework developed throughout this chapter. Potential licensees are always seeking ways to avoid entering into licenses since designing around a narrow patent can be easier. Licensees may also claim the patent is unenforceable due to inequitable conduct before the PTO. These inequitable conduct claims are not limited to clear cases of fraud before the PTO, but can include oversights such as the failure to cite relevant prior art known to the patentee during prosecution of an application. Prudent businesses practices and the use of experienced patent professionals can prevent such losses.

CONCLUSION

This chapter provides a survey of some of the important aspects of an aggressive IP strategy. Strong IP portfolio development and profit maximization is a continuing process requiring vigilance and an appreciation of IP's importance. To this end, utilization of forward growth orientated business practices incorporating aggressive IP strategies can provide a solid cornerstone for the success of any company.

Market Based Business Development
Ryan Bethencourt

Ryan Bethencourt is a senior business development professional and has worked for both U.S. and European biotechnology and pharmaceutical companies in a variety of business development roles. He has also co-founded several start-ups. Ryan has a master's in Bioscience Enterprise (an MBA/Biotech fusion course) from Cambridge University and can be contacted at *R.Bethencourt.02@cantab.net*.

Business development is a vague term, and is used to describe a large variety of business roles within the broader healthcare industry. I define biotechnology/pharmaceutical business development as a mixture of scientific, legal, financial, deal making, and selling roles. While accurate, this definition does not clarify the role itself. No two business development roles are alike and anyone involved in business development will likely need to use their resourcefulness, creativity, awareness, and full range of business and scientific skill set to be successful. Many business development roles combine both the buy-side and sell-side responsibilities involved in developing the businesses products. Even within some of the more academic licensing-only roles, there is a fair amount of exposure to selling either concepts or ideas to potential partners.

In this chapter I describe my insight into business development from the perspective of looking for a deal. I provide suggestions on analysing markets, targeting compounds/companies, and determining which opportunities would most likely fit into your pipeline. Business development is still very much an art, and is sometimes far removed from science; although there are some great financial models, they require a static market place that does not exist.

Great business developers look at large and chaotic market environments and call the right shots (which can only come from experience) through a mixture of extensive research and "gut" feelings.

IT'S A CROWDED MARKET; WHERE DO YOU WANT TO SET UP YOUR STALL?

Most healthcare companies have mission statements which read:

> "We develop industry leading medications for the treatment of [INSERT INDICATIONS] and are committed to being a global healthcare leader and changing the lives of millions of people through access to our safe and effective medications."

These are fairly standard and bland corporate mission statements, and they provide very little insight to those trying to look into the future of dramatically changing global healthcare systems due to a confusing mix of development, regulatory, and payer hurdles. Interestingly enough, and increasingly, these mission statements are seen as useless by Wall Street and the City of London. Active, focused mission statements are increasingly important.

Large pharmaceutical companies' market capitalisations have decreased over the past couple of years. The reasons for these decreases are related to an increase in generic drug competition, speculations on changing business models, and general research effectiveness and development spending. Companies must sharpen their focus in identifying which particular areas of unmet need are ideally suited for a particular company.

So which specific therapeutic indication is most suitable for your company: CNS, respiratory, oncology, RNAi–the new new thing? This is very tricky ground as sometimes the most exciting compounds can fail at different points in a product's life cycle. Pfizer's Torcetrapib was an extremely exciting product with the potential for raising good cholesterol and lowering cardiovascular risk. The drug failed during clinical testing. Merck's highly successful osteoarthritis drug, Vioxx, was commercially successful until fears were raised about its potential link to increased cardiovascular risk. So even when you pick a disease area with an established and large potential market, the unknowns of human medicine can work against a healthcare company's best intentions.

Assessing a market:
1. Identify whether your company has the required capabilities
   - Can your company effectively leverage the product? It is unlikely you will be able to effectively compete with a big pharmaceutical firm selling a primary care product if you are a small biotechnology company.
   - Can you afford to compete for compounds in this arena? Looking at the recent deal sizes in that therapeutic category, would your company be able to afford an acquisition/licensing deal in this area?
   - Is the clinical development pathway fairly well known or is there likely to be many unknowns due to new mechanisms of action of the new chemical entity (NCE)?
2. Picking the Market: Data, experience and intuition
   - How large is the potential market?
   - What does the future regulatory and reimbursement environment look like? One particularly relevant therapeutic indication facing reimbursement challenges in the future is oncology. An increasing number of new therapies are being developed for a wide range of cancers. This progress is great for patients. However, average dosings of these oncology therapies is expensive. Some regimens cost several thousand to tens of thousands of dollars. This is disconcerting for both payers and politicians. Payers are upset these therapies have unwanted effects in their budgets. From a political perspective, increasing costs mean some individuals are unable to access the latest therapies. This scenario is particularly

true for those patients in countries with socialised, and hence rationed, healthcare systems.
- Will you have a limited product launch window? Be very careful, as delays in clinical developments often occur. A tight product launch window could affect your company's potential to compete in a specific market, and therefore affect this product's potential value to the overall pipeline.
- Is your product serving a large unmet need rather than an incremental improvement on a current therapy? Serving existing markets can be lucrative if you feel a minor improvement would have a substantial effect on a potential market, but you may also be affected by future changes in reimbursement for non-novel products.
- Have you talked to those involved in the market to assess whether they feel there would be a real need for this product? Would patients find this NCE or new delivery mechanism useful? Many companies are blinded by the potential data behind a product. Sometimes companies discover patients won't use their product for seemingly trivial issues. The asthma market is an example of the need for early feedback. Asthma products succeed or fail depending on perceived ease of use of a particular inhaler.
- What geographical market do you wish to tackle? Major healthcare markets are very different and can have substantial differences in medical treatment protocols. Cultural differences can also prevent success in one geographic market translating to another. Suppositories in France are popular for many conditions including pain relievers. In the United Kingdom, however, suppositories are culturally unpopular.
- Are there enough products in development? Are

you willing to identify candidates in early stages of clinical development and invest earlier rather than at a later, less risky phase?

THE HUNT IS ON!

Once you've clearly identified the therapeutic market which best fits your organisation and provides the necessary scope for success, you've got to start filtering your potential target list. This may sound easier than it actually is. Most biotechnology companies list products in clinical development on their websites. Third party providers can provide market briefing documents, therapeutic overviews, and provide their own forecasts. Interestingly enough, most experienced business developers do take notice of these extensive proprietary information sources. Developers tend to use them as secondary data sources, preferring to piece together their own tailored lists. The process of filtering through both public and proprietary data to compile a top-20 or even a top-100 list of global candidates is very laborious, but is essential in developing a marketplace understanding and in projecting future market developments.

This is a pattern recognition exercise. The deeper you get into the specifics of individual products and the essential bits surrounding due diligence, the more you can start filtering and building your own mental model of the future marketplace. The art of finding a potential target opportunity is in analysing the market and recognizing patterns. While sophisticated financial models can be used to augment the decision making process and proper due diligence provided by medical, legal, and business experts is necessary, the industry ultimately relies (surprisingly) on mavericks willing to make confident decisions with a large number of unknowns and a high level of risk. Postponing difficult decision making leads to the paralysis sometimes seen in larger pharmaceutical companies. They attempt to de-risk the drug development process.

SPECIFIC ISSUES WHEN ANALYSING POTENTIAL TARGET COMPANIES

Most biotechnology companies are valued based on their growth prospects and their products' potential market. This can be used as a guide to identify how well a particular biotechnology company's products would fit into your company's pipeline. Searching through publicly available news feeds is one way of assessing obvious difficulties arising with either the company, the quality of their science, or any problems with their product development. Sometimes adverse news feeds can make a company more willing to negotiate, but be wary of this approach as it should be used with caution.

Gauging the financial health of a prospective partner is crucial. Ensuring the partner is sufficiently capitalised to continue clinical development, has sufficient income (if applicable), and is meeting their clinical development milestones, either with your target product, or with others within their pipeline, is important.

One of the simplest ways of evaluating of a large number of companies is employing SWOT analysis (strengths, weaknesses, opportunities, and threats). This analysis is useful in providing a general overview of a product or company and can be used as a discussion document with your team. Dedicate no more than one page per opportunity to ensure these documents are used as quick references. There is also a large variety of mathematical tools which are used in business schools to analyse opportunities. These include net present value models (NPV), internal rates of return (IRR), and more complex financial models.

A financial model is only as accurate as its weakest assumption. Hence, I recommend using the simplest forms of a mathematical analysis to ensure a deal makes sense financially for both parties. A complex analysis could inappropriately measure the "fit" of an opportunity.

How do you move forward once you've identified your top-20 target companies or compounds? This is where the selling begins. One has to begin contacting these companies either directly (which can include direct calls, networking events, and medical/

scientific conferences), through a third party (perhaps an investment bank), or through relationships held by individuals within your own company. It is very important to be clear on how you wish this process to progress before this process begins. Would your company like to proceed in a discreet way or would you be comfortable if there were a higher level of public awareness of your interest in potential acquisitions? Most companies prefer discretion and will insist on signing a confidential disclosure agreement (CDA) before discussing any details. This document can be provided by a specialist pharmaceutical/biotechnology lawyer. CDAs should normally be a quick formality. Sometimes, however, CDAs have unusual requests requiring further time to negotiate. The process of targeting and qualifying potential companies can take a surprisingly long amount of time. Be ready for this and make sure to set expectations internally within your department and with senior management.

How do you qualify targets and pull together a final list of three to five potential partners/acquisitions?
1. Medical/scientific due diligence
2. Validation for the potential market
3. Personal chemistry

Personal chemistry is particularly important, as the other points are standard business considerations. It is surprising how many deals can fail primarily because the chemistry between two companies or teams just does not work. A product may be a great fit, but if the chemistry of the team is not right, then there are likely to be downstream problems in making a deal. If, for whatever reason, a great company with a great product does not feel right, it may be worth walking away.

There is an unlimited range of options available when making a decision on the type of offer to make once you and senior management have agree to proceed with a product. In general, offers fall under three categories: a co-promotion deal (usually for later stage compounds and more developed partner companies);

full product acquisition (either from clinical development or marketing authorisation onwards); or, acquiring the target company outright. Each strategy has its own advantages. Potential pitfalls are the subjects of whole books. Including a small, senior team of key decision makers helps ensure the deal making can occur with minimal internal resistance. There will undoubtedly be delays and unexpected problems, but having a team of key internal opinion leaders will help buffer these potential problems.

THE FINAL, LONG STEP — GETTING TO A SIGNED CONTRACT!

So assuming that all stakeholders and future partners have agreed to the initial deal, the final step is to iron out the details of a written agreement. As I'm sure most of you are aware this will likely be a long drawn out process and one that can be extremely unpredictable. I have known of deals that have taken literally years to conduct. Yet, there have been others that have been agreed literally over night. The most rapid deals tend to be those where there have been established relationships, between individuals that have had high levels of interpersonal trust established from previous working relationships.

Due diligence is absolutely crucial and most established Healthcare companies would not allow any potential partnership/acquisition to proceed without it. One important point to note is that even if acquisition/partnership talks are being held in confidentially, it is likely that other companies are proceeding in a similar fashion. If so, you and your firm have to be very clear from the start on how you would either compete in a potential bidding war (i.e. ensure that you have a clear strategy your senior management team have approved to move rapidly) and how best to minimise the risk of competition. It's important that if your organisation wants to negotiate exclusively once a verbal commitment has been reached that the method of retaining this exclusivity is put into place rapidly.

In almost all cases the final agreement will be dependent on the necessary legal, business and scientific due diligence being conducted. This is crucial as it is important to establish the accuracy of the information on which the company or therapeutic has been sold on. Many compounds in clinical development have been subject to multiple pull through agreements and may have unworkable complexities built into the extensive legal agreements surrounding their development. One such complexity may be the need to pay third parties royalties based up certain sales milestones, which would detract from the value of the compound. It may be in certain cases that the intellectual property behind some compounds may not have been as watertight as either you or your potential future partner believed.

The final agreement will require an extensive amount of input from a broad spectrum of internal and external experts experienced in specific therapeutic areas, regulatory development and the legal intricacies of therapeutic deals. There are specialists lawyers, consultants and investment banks that tend to be clustered around major biotech/pharma locations that can provide these services. They will be able to assist you in outlining a variety of legal/business scenarios and depending on the nature of the transaction, licensing or acquisition, will outline possible contract options. From my own experience, in particular, legal reviews can be challenging for the individuals involved as it can be an arduous process (even when both parties just want to get the deal done).

Some agreements—at least from a legal perspective—can often be viewed as the least worst option to make a deal happen. No deal is perfect and once you accept that principle it makes discussions around certain clauses more bearable. It goes without saying that a full review should be conducted by qualified members of staff and even in smaller companies, cutting back on legal or regulatory review due to the cost implications would leave a large potential liability. Make sure you also personally review the entire contract and that it all makes sense to you, lawyers can provide put together excellent contracts but you and your counterpart are responsible for the operational mechanics of how to move things forward once the final signature has been placed on the contract.

Lastly, I'd like to add my own thoughts in regards to making a successful deal. Successful deals are all about trust and relationships, our industry is a surprisingly interconnected one and the most successful companies (apart from having large cash reserves) are those that negotiate deals that are fair on both sides. Just to end with one of my favourite Chinese proverbs that reflect the current challenges we have within the pharmaceutical/biotech industry "To open a shop is easy, to keep it open is an art."

The Ins and Outs of In- and Out-Licensing
Gil Ben-Menachem[1]

Gil Ben-Menachem is a Director of Business Development at Paramount BioSciences, where he in-licenses promising drug candidates. Prior to joining Paramount, he was at the National Institutes of Health, Office of Technology Transfer, where he out-licensed numerous NIH Technologies to biotechnology and pharmaceutical companies in the US and abroad. At the National Institute of Child Health he led the development of a novel glycolipid-based vaccine for Lyme disease. He was a Co-founder and Chair of the NIH Bio Science Business Interest Group. Gil holds a master's in Biotechnology and a Ph.D. in Microbiology from the Hebrew University in Jerusalem, and an MBA from the University of Maryland. He is author of over a dozen scientific and business papers, and the inventor of several patents. He can be contacted at ben-menachem@gil.com.

Licensing, whether in- or out-, is an inherent activity of every biotechnology company. Rarely does a single company advance a drug candidate from discovery to approval on its own. Rather, drug candidates exchange hands on a daily basis among academia, startups, biotech and pharmaceutical companies. Understanding the process, the driving forces, and the people behind it, are crucial in a successful licensing strategy that can make or break a company.

In this chapter I share my experience working on both sides of the negotiating table: out-licensing at the NIH Office of Technology Transfer, and in-licensing at Paramount BioSciences. I will briefly describe the licensing executives I have encountered, their role within the organization, the initial steps of the licensing deal, the scientific and business evaluations, and the due diligence process and negotiations that lead to a successful partnership.

---

1   *Opinions presented are those of the author and do not represent the views of Paramount Biosciences or its Affiliates.*

THE BIOTECH BUSINESS DEVELOPMENT EXECUTIVE: BACKGROUND, TRAINING, EXPERIENCE AND DESIRED PERSONALITY TRAITS

Understanding the professional background and education of the biotechnology business development (BBD) executive is important not only to become one, but also to fully appreciate their level of scientific and business experience. The cliché that the first impression is the most important one is also very true in biotechnology licensing. It is crucial to understand with whom you are doing business. Does he or she have a PhD? MBA? Was he or she trained as a lawyer? For which companies did he or she previously work? Browsing the management bios on a company's website, searching PubMed for scientific publications, or performing a Google search will provide you with critical background information and topics for your first conversation.

Unlike business development (BD) positions in other industries, scientific background and education is an essential part of every BBD executive's training. Most people with BD functions at biotechnology companies have a MSc, PhD, or MD, which is often accompanied by a MBA or sometimes a JD In many situations these letters translate into salary numbers. In most cases the scientific training precedes the business training. The reverse route of first obtaining the business education and then completing a graduate program in the life sciences is not very common, for what MBA graduate, with six-figure earning potential would pursue an additional four to six years of laboratory work with a postdoc's salary? One needs to have a deep understanding of the company's scientific principles, technology platform, specialty products, and development plan. A first-year biology undergraduate course will not suffice, nor will the average MBA curriculum.

Many of the biotechnology executives I meet started their business careers by commercializing their own scientific work. This was accomplished by patenting their inventions or commercializing work conducted by their supervisors and colleagues.

The transition from the "bench to the boardroom" can be a straightforward process when the company is a young startup and the BD function is filled by a former colleague or student of the founder. Sometimes the process can be long and require many years of experience before the executive ends up in the BD role of a more established firm or a large pharmaceutical company.

Often, a position with an academic technology transfer office serves as a "prep school" for a future BBD position. Academic licensing positions are in many ways similar—yet different—to a company BD position. One major distinction between these two positions was nicely phrased by a former colleague; "A BD position is very much like a licensing position, only your job does not depend on the deal." Obviously, she was on the technology transfer side. There is a lot of truth to that statement. However, the fact that one's job does not depend on the deal (which comes down to evaluation criteria—is the BBD measured by the number of deals closed, or by the number of patents filed?) has many implications concerning the differences between an academic tech transfer position and a BD biotech company position. The subject merits a chapter of its own.

Former associates and principals at venture capital firms often end up filling the BD role of companies they finance. I have come across numerous BBD executives who, prior to their full-time BD position, were involved with the company through their position at a venture capital firm. Often this connection is initiated by a member of the board of directors. In most cases the BD executive has very intimate knowledge of the company and its business and possesses an insider's perspective of the company and its BD needs. This is a natural career route for venture capitalists for whom becoming a venture partner is not an option. This transition allows BD executives to utilize their network of contacts within other biotech companies. The Investment community or Wall Street contacts are also valuable especially when an IPO, PIPE, or other types of investments are pursued.

BD executives are usually nice individuals. This generalization sounds trivial, but if you think about it, this characteristic is in the essence of the profession. Every licensing deal, no matter how big or small, comes down to basic trust between two parties.

Both sides are represented by licensing executives, and trust must be established by BD executives. The licensor's proprietary technologies, drug candidates, intellectual property, and the licensee's financial considerations (e.g. upfront payments, milestones, royalties) and commitment to develop the assets will eventually result in a licensing deal only if the two parties can communicate and trust each other. This exchange is carried out "belly to belly." That is, of course, before valuations and lawyers come into the picture. In order to build trust, the company representative or lead role must be trustworthy, reliable, open, and honest. It is very difficult to prosper in this line of business if one's demeanor is not social, but instead is condescending and arrogant.

The characteristics that make a good BD person are numerous. Among these are honesty, organization, structure, an open mind, and being a team player. Another particular feature is motivation. A person's strong desire to close the deal will significantly speed the process and increase the chances the deal will close. On the other hand, an executive who is not motivated will not push forward strongly, and can jeopardize the chances of the deal closing. In the end it all comes down to incentives. A "well-fed" content executive whose performance is not measured by successfully closing the deal, and whose job does not depend on the deal, may end up sabotaging and/or jeopardizing it.

THE NETWORK: THE MEANS TO AN END

People tend to overemphasize the importance of one's network of contacts. It is often said that a CEO or a BD executive's value is measured by their Rolodex/Blackberry contact network. I disagree. While there is no doubt that playing golf on a regular basis with Pfizer's Chief Licensing Officer is fantastic, a good BD executive is measured by their ability to identify the target and go after it, and, in many cases, to pursue the target in the absence of personal connections. As a director of business development, my job is to prospect promising drug candidates, to initiate contact with the potential partner and to develop a thorough understanding of the opportunity both in terms of "classical" due diligence (i.e. scientific background, medical need, intellectual property, potential market size), as well as deal assessment. Deal assessment means understanding, in the very early stages of negotiation, if the business terms for both sides will fall into place.

The specific connections needed to identify a drug candidate for possible in-licensing (e.g. target indication, technology type, development stage, financial expectations, etc.) can seldom be met solely by tapping one's established network of contacts. The existing network is merely a starting point and should not be regarded as the only tool, or the main tool, in accomplishing the task.

Of the recent licensing deals I negotiated, only one originated from a previous acquaintance. Both my colleagues and I have years of experience and have met and worked with numerous industry leaders. Yet a majority of the deals we close result from actively pursuing the leads we identify, and are further pushed through cold calling. On one occasion, we interviewed an executive who, during the interview, bragged she had more than 4,000 contacts from the biotechnology industry. Other than demonstrating this particular individual's obsession in cataloging the business cards she received, her network served no use in demonstrating her abilities as a good BD executive.

Another way to think of a contact network is as a means to an end—and not the end itself—similar to a PhD program. The fact that a BD executive has a PhD doesn't mean he or she is necessarily smarter than a BD executive having only completed a bachelor's or master's degree. Rather, a PhD demonstrates that he or she knows a lot about a very specific, limited niche of subject matter. What does that have to do with accomplishing BD goals? Nothing. However, the assumption is that someone who completed their PhD learned how to learn. Four or five years spent in a lab conducting their own research, setting goals, and acquiring deep knowledge in an area that was previously unfamiliar demonstrates an ability to learn independently. Likewise, a successful BD executive expands a network of contacts through successful interactions with other key people in the industry. The executive must "learn" how to expand their network. The learning process leading to this success doesn't come without effort. The confidence to target any executive and the ability to do so is much more important than knowing this individual *a priori*.

Websites such as LinkedIn, which make use of the *six degrees of separation* concept (any individual can be reached through a maximum of six nodes of a friend who knows a friend who knows a friend...), have some utility in maintaining and expanding networks, but are better suited for job searches than for actual business transactions or lead identification.

Some business schools offer master's-level courses in networking and teach its focus on importance in career development. Good networking skills and the ability to reach out to any company or individual is not something learned in a classroom. Rather, it requires adopting the right "state of mind" and having an experience-based appreciation of the process. For example, as a new postdoctoral fellow at the NIH, I needed a way to get acquainted with the prosperous Mid-Atlantic biotech industry. Looking within the NIH for connections to the region's biotechnology leadership, I realized that other than the Technology Transfer Office, there was no formal organization interacting with local venture capitalists and biotechnology companies. I decided to co-found a special interest/discussion group at NIH, the Bioscience Business Interest Group (BBIG). We held monthly meetings and hosted speakers from the local community to lead informal discussions about science, business, and entrepreneurship. Within months, BBIG expanded to over 250 people. This strategy allowed me to get to know some very interesting scientists, CEOs, VCs, and entrepreneurs. I made many new friends and contacts in a relatively short period of time.

To conclude the networking section, it is crucial to understand that while it is important to expand and maintain a large network of contacts, this should not be regarded as the sole goal. Developing good interpersonal skills, being able to approach any target and presenting your case in an appealing way, is equally important.

BUSINESS DEVELOPMENT: WORKING UP, DOWN, AND ACROSS THE ORGANIZATION

The BD role is unique within the biotechnology company. The business development function is located at the junction of the company's core assets—the science, the business model, the people, and the intellectual property. Unlike licensing positions in a technology transfer office or venture capital positions where BD executives have to understand relatively little about many fields and technologies, the biotechnology company BD executive has to demonstrate deep knowledge in a very specific field which revolves around the company's assets. He or she has to know everything about the company—the management, the science behind the compounds or platform, the clinical development path, regulatory strategy, manufacturing process, market positioning, investor perspective, and obviously the company's financial position and projections. The BD executive must also know the market, the players in the specific field, the competitors, and the assets being developed. In order to obtain this understanding, the BD executive has to interact closely with all departments and individuals within the company. He or she has to report to top management, obtain information from the research group, the legal department, the regulatory group and manufacturing, and integrate all this information into a coherent body of knowledge that can be shared with internal and external entities.

On one occasion I asked a company's BD director a question regarding a drug candidate we were evaluating for in-licensing. The question involved the company's positioning strategy given a specific competitor who was developing a similar drug. I emailed the question Saturday night around 9 p.m. and expected an answer at the beginning of the following week. Amazingly enough I received an answer 30 minutes later, typed from a Blackberry. The message was a very thorough (300 word) analysis of their compound's advantages and how it differed from the competitor. There is no way the BD director could have consulted internally or had access to a pre-existing document on such a short notice. He wrote it on the spot. Obviously, I was very impressed. This example demonstrates the knowledge level desireable for out-licensing activities. The BD role is truly a cross-functional matrix-oriented job. When done well, it allows one to enjoy all areas and functions of a biotechnology company.

Likewise, it is important the BD executive has sufficient autonomy when dealing with in and out-licensing transactions. I have encountered numerous BD executives who did not have enough authority to engage in meaningful business discussions, which slowed the negotiation process and created unnecessary tension between the parties. Obviously the board and the CEO have the final say, but the BD executive should aspire to create a situation where he or she understands, appreciates, and can represent the company's strategy when it comes to licensing. Speed of negotiations is always a key factor in evaluating the prospects of a deal. Requiring supervisor approval at every step along the way can wear the parties out, and can be detrimental to transaction momentum. In one situation, I worked for several months with a BD director to in-license a promising drug candidate. We assumed he had mandate to assist us in the due diligence process, the term sheet negotiations, and the closing. As it turned out, the CEO was not in the loop during the entire process and ultimately killed the deal. We were amazed to see the deal on which we worked so hard crumble in minutes at the execution stage. This was a complete waste of time. If the CEO had been informed of our negotiations prior to being presented with the signing papers, we would have known the deal wouldn't work.

FIRST CONTACT: PREPARATION, CULTURAL DIFFERENCES AND FIRST IMPRESSIONS

Do your homework! Prepare by thoroughly exploring both the company and its team members before initiating contact with any company or BD executive. The company's website, press releases, SEC filings, proprietary and public databases, as well as a simple web search on the company, its drug candidates, or the people behind it, is essential information every BD professional needs to gather prior to initiating contact. Other pay-per-use proprietary databases and professional journals are also great resources. Preparation can make the difference between a good or bad impression in an initial contact, which is often the most important and hard to change. Likewise, examining any previous discussions your organization has had with the prospective individual or company is equally important. Proper preparation should be standard operating procedure in every BD role.

Cultural differences between the licensor and licensee also play a significant role in the deal flow process. Bridging these differences is usually the role of the BD executive. Every international business transaction should take cultural traditions and practices into consideration. In Israel, for example, oral communication is very direct and sometimes blunt. It took me a while to become acquainted with the more subtle American style. I was unaware that this cultural difference could sometimes be perceived by my American colleagues as rude and impolite. Likewise, when my American supervisors said, "you should consider doing…" I interpreted it literally, as a mere suggestion, rather than a directive. In Israel a boss would usually say, "Do this!"

There are numerous resources available for explaining and studying cultural differences in order to avoid conflicts. The way of doing business is slightly different in various communities, and it always helps to know whether a Japanese counterpart's "yes" means really "yes," or "maybe," or "yes, but six months from now," or "I don't know, I'll have to ask my supervisor." Very seldom will you hear a direct "no."

DEAL BASICS: THE SIX W'S

Once initial contact has been established and there is a basic understanding of mutual interest in pursuing a licensing deal, the next step should be to delineate each party's requirements and expectations. I often call it the "Six W's," similar to the four "P's" of marketing. The six W's are: Who Wants What, Why, When, and Wow (how much?).

Who Wants What

The most common agreement is between two parties: the licensor and the licensee. However, sometimes there are other parties directly or indirectly involved in the deal. Clear identification of who owns the technology is not always obvious. When referring to the "technology" I am including various assets such as the assignee of the patents, know-how, trade secrets, and manufacturing (CMC). I have come across licensing opportunities in which these various components were not owned or controlled by a single entity (the licensor), which can make negotiations difficult for the licensee. Sublicensing is a major topic in these situations, and full understanding of licensing agreements preceding the current transaction, as well as potential future sublicensing agreements, is of the utmost importance.

The *what*, as I indicated earlier, can be problematic. In addition to questions regarding the specific technology as discussed above, patents, know-how, and trade secrets are questions related to the field-of-use. These topics include multiple indications (drugs that can be developed for more than one disease or patient group), method of administration (for the same or different indications), and geographic locations (*territory*). Is the technology a platform? What happens with future compounds? Who owns future intellectual property? These critical issues are all decided and agreed upon in the licensing agreement. Nevertheless, it is wise to try and address these topics at the initial stages of the discussions to prevent unbridgeable gaps later on.

Why

It is important to consider motivations. Why does the licensor want to out-license the technology? Also, why does the licensee want to in-license? Has a lack of sufficient resources like capital, knowledge, or time forced the licensor into considering a partnership? Perhaps the licensor does not believe in the technology's potential? The licensor's main objectives in the partnership are often referred to as "WIIFM," or "What's in it for me?" Are the upfront payments requested by the licensor to maximize the immediate cash-on-hand? Is the licensor partnering to see the technology developed more rapidly? This arrangement is often referred to as "back-end-loaded" versus "front-end-loaded" deals.

As for the licensee, what are their main interests? Does he or she seek a true partnership or full control? Will he or she immediately "flip" the technology and sublicense it without further additional development? These are all examples of essential questions every BD executive should ask themselves when engaging in a licensing transaction. The sooner he or she obtains answers to these questions, the better. It is extremely frustrating to come across unpleasant "surprises" that could have been avoided much earlier in the process if the right questions were asked. The BD executive must ask these questions in the preliminary stage of the transaction.

I believe every licensing deal has to be a win-win situation. There will not be a deal if both parties have nothing to gain. Early understanding of the true *why*, goes beyond the obvious "why does the other side want to do a deal?" and extends to deeper issues such as "why are they pushing back so hard on this point?" Identifying the answers to questions like "Why does it take them so long to make a decision?" and "Why is this point so important to them?" will result in better appreciation of the needs and wants of the other side. Also, accommodating these issues increases the chances of successfully completing the deal.

When

When is an important question in the licensing process. When will the license terminate? Expiration of some licenses is invoked when the patent expires. Other definitions are also common. The license should provide exact timing of initiation, payments, and milestones. These issues become more complicated when there are co-promotion, co-marketing, or co-development agreements at stake.

Wow (How much?)

There is no doubt this consideration is one of the most important factors in any licensing deal. The models used to determine the net present value of a drug candidate, project, or company, and the assumptions used for these calculations, are beyond the scope of this chapter. Bogdan and Villiger[2] provide a great resource. However, my experience on both sides of negotiations is that there is a lot of leeway for both licensor and licensee when calculating valuations. Non-monetary factors are often more important. From the licensor's point of view, questions such as, "Does the licensee have the knowledge to further develop the technology? Do they have the necessary financial resources to do it? How aggressively will they develop the technology? Will they 'shelf' it?" are as critical as the upfront milestones and royalties on the table.

The NIH Office of Technology Transfer assumes a policy that future development of the technology and its potential to benefit the public (especially Third World countries, and public health non-profit organizations) is more important than immediate compensation and licensing fees. From the licensor's point of view, non-monetary considerations include the ability to further develop the technology, reliance on the licensor to smoothly and quickly transfer the body of knowledge accumulated on the technology, and the validity and substance of previous work carried out by the licensor. One relevant example involved a small biotech company trying to out-license a drug. The drug's owners claimed the drug was ready to enter Phase II clinical trials for cardiovascular disease. Only in the very late stages of the discussion did I understand that the protocols and results for this drug candidate would not be available for the big pharmaceutical company attempting to in-license the technology. The preclinical work, IND submission, and Phase I Study would have to be repeated. As a result, the valuation of the technology decreased significantly and we decided not to continue discussions due to reliability issues with the potential licensor.

---

2   Bogdan, B., Villiger, R. *Valuation in life sciences*. Springer-Varlag Berlin, Heidelberg, 2007.

BUSINESS AND SCIENTIFIC DUE DILIGENCE

The due diligence process can be easily facilitated by understanding the other side's needs and wants. Among the hundreds of technologies I have evaluated, I have seen data packages that were pleasurable to read. Conversely, some packages were so poorly written that I didn't even share them with my colleagues. The first exchange of information between a licensee and a licensor is usually a non-confidential package describing the technology, method of action, initial results, team, intellectual property, and potential market size. An executive summary and business plan are often included. Once initial interest has been established, more complete and detailed data are exchanged after executing a confidentiality agreement.

The licensor needs to be open and receptive to various inquiries, and willing to provide any required information. I have come across technologies, specifically in tech transfer offices, and could not gain access to the inventors to ask for further information beyond that provided by their tech transfer office. In many cases such resistance is indicative of the down-the-line process and may be detrimental to advancing discussions. Likewise, delays in reaching resolution in the confidentiality agreement terms may signal a warning of what to expect when crucial negotiations take place regarding the term sheet and licensing agreement.

One of the most frustrating scenarios is when the licensor needs to "milk" or "extract" information from the licensee. The licensee can be resistant to sharing information that was difficult to produce. However, a coherent and responsive approach is essential for the process to proceed and run smoothly. Withholding information on a need-to-know basis impedes the due diligence review and wastes precious time when complicated partnerships are at stake. It is beyond the scope of this chapter to delve into IP due diligence, scientific due diligence, and valuation due diligence. These topics are usually conducted by specific business units and by different individuals or groups. The BD unit has the responsibility to integrate all the information and move the process forward.

The due diligence process can be very stressful for the licensor. It will require availability from all team members who were involved with the technology development. External regulatory, manufacturing, clinical, and preclinical consultants may also need to be included. Often third parties are associated with the process, such as CROs, manufacturers, regulatory consultants and outside counsel. The BD unit will coordinate these different entities and facilitate the process by addressing each party's needs.

This essential part of the licensing transaction is the longest and most difficult stage. It may last for an extended period. Nevertheless, when done properly, the due diligence process will enable a smooth transaction and better understanding of the technology by the licensor. Upfront diligence will save a lot of time when the transaction is completed and the licensor takes over the technology development.

Intellectual property due diligence can also be lengthy and expensive. The licensor needs to fully understand the technology's patent protection, expiration dates, potential for off-label competition, and freedom to operate. External counsel is often sought, adding to the costs.

When checking the "freedom to operate" for a diabetic drug, I discovered that the patents were valid and the claims covered its use. I also discovered an article appearing fifteen years earlier in a Brazilian journal describing the same compound. This can be considered prior art. The fact that the United States Patent and Trademark Office (USPTO) grants a patent does not necessarily mean there is freedom to operate. No business would want their patents invalidated (see also Yeda vs. Imclone[3]) once their drug is on the market generating substantial income.

---

3   *United States District Court, Southern District of New York. Yeda Research and Development Company, Ltd. vs. Imclone Systems Inc. and Aventis Pharmaceuticals, Inc. Retrieved electronically from http://www.nysd.uscourts.gov/rulings/03CV08484_opinion_091806.pdf*

VALUATION

The valuation stage is a critical element in the licensing transaction. Parties reaching this stage have conducted prior due diligence steps (scientific, IP), and are now ready to place a dollar value on the assets. Reviewing the entire process of valuating an asset is beyond the scope of this chapter, but net present value (NPV) and the risk-adjusted NPV (raNPV) are the most common models used in asset evaluation. The reader is referred to Bogdan and Villiger[4] which describes these models and the proper way to implement them.

There is a delicate interplay between number crunching, experience, and instinct when valuating a project or a drug candidate. I am amazed at the significant part instinct plays in this process. While the science behind a drug candidate needs to be rock-solid and backed by numerous experiments, publications, and solid proof of method, when it comes to putting a dollar value on the assets, the "art" is sometimes as important as the science. There are many factors here, not all of them monetary. Obviously, the licensor seeks adequate value for the assets and wants to maximize the deal. However, non-monetary offers can have substantial value for the licensor, as well as for the licensee.

Listed below are key considerations in determining the value of an asset:

Development Stage
- Bringing an advanced drug to market costs less and reduces risk. The raNPV model takes into account the probability that a drug candidate will advance to the next stage using a decision tree method.

Indication
- The indication or disease group will have significant effect on the success rate and on future development costs.

Peak sales
- Peak sales is one of the key factors affecting the

---

4   Bogdan, B., Villiger, R. *Valuation in life sciences*. Springer-Varlag Berlin, Heidelberg, 2007.

NPV valuation of an asset.
Cost of Capital
- Cost of capital, or discount rate, has a significant impact on the value assigned for the asset.

Less-quantifiable value drivers include:
Dedication
- How dedicated will the licensee be when it comes to developing this asset? Is the development plan adequate? Will it be followed? This is basically a question of trust.

Team
- Is the team responsible for the development capable and experienced in developing these or similar assets?

Publicity and Marketing
- How will the deal impact the parties in terms of public relations, stock price, analysts, shareholders, investors and other stakeholders?

Time to Closing
- How fast can a deal be executed? What resources will have to be allocated to the transaction, to the partnership, and at what costs to the organization?

Some of the recent licensing deals I closed were completed in as little as nine weeks. After the deal was signed and sealed, it became clear that the speed of the transaction was one of the most important factors to the licensor, offsetting a potential higher deal value which may have taken much longer to achieve.

NEGOTIATIONS

Ideally, the licensor and the licensee should arrive at the negotiating table after they have reached a basic understanding of the scientific and clinical data, intellectual property position, valuations, and the basic deal terms.

Often, parties begin negotiations while evaluation is still underway. In fact, some types of negotiations should occur as early as the first contact, in order to spare the parties' time in discussions. It is much better to "kill" a drug candidate early in the development stage, when relatively low costs have been incurred, than to discover only later, after millions of dollars have been sunk into preclinical and clinical trials, that the drug candidate doesn't show efficacy. Likewise, BD executives should aspire to "kill" the opportunity as early as possible, avoiding a waste of time and resources for themselves and their counterparts. Nevertheless, every chance of moving the deal forward should be fully explored.

Entering formal negotiations too early may harm the prospects of the deal due to a lack of basic understanding, trust, and appreciation for the other side's needs and wants. In one situation, we concluded a negotiation and reached a final licensing agreement before the entire team had a chance to weigh in on the development path. Misunderstandings emerged regarding the actual development stage for the drug and what would still be required to submit an IND. We had to go back to the table and renegotiate with an entirely different valuation analysis, which was a very tough position.

On another occasion, we reached the negotiation table too late in the process. While the entire team was up to date and we concluded the due diligence, the due diligence took too long (several months). The parties were exhausted after many back-and-forth questions and answers, conference calls, and trans-Atlantic travel for face-to-face meetings. When actual negotiations began, the expectations were so misaligned that the exchanged term sheets were several orders of magnitude apart. This could have been avoided had we entered the negotiations earlier.

Every BD executive should become acquainted with concepts such as "Best Alternative to Negotiated Agreement" (BATNA) and WIIFM through the general literature.[5] There are three general rules of thumb worth mentioning when discussing negotiations. While very basic and straightforward, I truly believe they

---

5   Fisher, R., Patton, B. M., Ury, William. *Getting to yes: negotiating agreement without giving in.* Penguin Books, 1991.

apply to every licensing deal.

1. "If there's a deal to have, let's have it. Otherwise let's not waste each other's time." While both parties have to be optimistic during the entire process, it also doesn't hurt to be realistic. If the process is taking longer than it should, the deal is probably not going to happen.
2. It has to be a "win-win," otherwise it will not happen. A "too good to be true" deal is bad for one party, or something is wrong.
3. Pick your battles. Make sure you know what the most important points and critical elements are, both for you as well as the other party. These are often not the financial considerations.

SUMMARY

Whether dealing with a simple material transfer agreement (MTA) between academic laboratories, or a $500M merger and acquisition with a major pharmaceutical company, the basic elements making both these transactions possible is portrayed in the licensing deal. In this chapter I have tried to describe the basic principles, the vocabulary, the process, and the people behind it.

I often feel the world of business development as it relates to biotechnology and pharmaceutical drug development can be described as a big, global marketplace (or a bazaar). The global marketplace is a sophisticated environment. The buyers and sellers are well-educated and savvy. The merchandise is very expensive and the expiration date is always an issue. There are sellers, buyers, regulators, financiers, and many other stakeholders. Mastering the art of bringing these parties together is not an easy task. In addition to knowledge, experience, and common sense, expertise in business development also requires understanding the ins and outs of in- and out-licensing. The big difference between this marketplace and any other, and the reason for my amazement and admiration of it, is that at the end of the day, the merchandise saves patients' lives.

Acknowledgments

I dearly thank Rosel Halle for her critical review, help, and friendship.

Free Cash Flow—the Essential Ingredient for Growing a Business

Gerald S. "Sandy" Graham

Gerald S. "Sandy" Graham provides coaching, advising, and consulting services on business planning, strategy, and development to university start-ups, new ventures, and small business enterprises. Sandy is a recipient of the Ewing Marion Kaufmann Foundation Internship, holds an MBA, an MS in economics, and has held senior-level positions in business and government with experience in the biotechnology, IT, telecom, financial services, automotive, agriculture, manufacturing, and game simulation markets segments. Sandy can be contacted at *s.graham4@earthlink.net*.

Capitalizing on the opportunity to develop therapies and cures for any of the prolific diseases, health, or medical conditions facing mankind has lead to a strategic partnership between science, medicine, engineering, and business in recent years. This collaboration paves the way for a proliferation of university-based biotechnology business startups engaging in drug and vaccine development, delivery systems, genomic profiling and engineering, cell biology, pathology, microarrays, proteomics, protein characterization, reagents and nanotechnology, as well as new software applications, medical devices, and instrumentation to support the growing biotechnology industry.

As university-based startups grow and expand into viable small business enterprises, an entrepreneurial dichotomy occurs: the urge and passion to focus on requisite technology and subsequent patents, and the necessity for following good business and management practices necessary in growing a new business. Biotechnology entrepreneurs have moved away from concentrating on SBIR/STTR phase I and II grants for R&D to commercialization efforts funded by angel financing and venture capital. Marketing and sales, investment in plant and equipment, human resources, as well as continued R&D is critical in making a business viable and establishing market share.

Employing strict financial management practices is every bit as important as developing the right therapies and cures. A strong cash flow picture, especially for the emerging small business enterprise, indicates sound business practices and attracts investment for expansion. In addition to the publication of typical financial statements like the income statement, the balance sheet, the cash flow statement and the statement of retained earnings, companies are also reporting on what is referred to as free cash flow. This statement reports key statistics which investors consider when deciding to invest. This paper will examine how to determine free cash flow using example financial statements, and will illustrate the importance of focusing on free cash flow as a means in determining the financial soundness of a small biotechnology enterprise.

WHAT IS FREE CASH FLOW?

The financial management text book definition of free cash flow (FCF) is "the cash flow actually available for distribution to investors after the company has made all the investments in fixed assets and working capital necessary to sustain ongoing operations".[1] In common terms, free cash flow can be seen as the level of remaining cash available to a company after all expenses have been paid. From the perspective of a young, small company, free cash flow is the essence and life-blood of the organization since it indicates the level of cash flow in excess of what is required to perform business operations.[2]

---

1 Eugene F. Brigham, Louis C. Gapenski and Michael C. Ehrhardt, "Chapter 2, Financial Statements, Cash Flow, and Taxes", pp 32-46; "Free Cash Flow", *Financial Management: Theory and Practice*, 9th Edition (The Dryden Press, Harcourt Brace College Publishers, 1999).

2 William W. Priest and Lindsay H. McClelland, "Free Cash Flow", *Free Cash Flow and Shareholder Yield: New Priorities for the Global Investor* (John Wiley & Sons, Inc., Hoboken, New Jersey, 2007): pp. 8-20, and George C. Christy, *Free Cash Flow: A Two-Hour Primer for Management and the Board* (Booklocker 2006): pp. 5-6.

The Securities Exchange Act of 1934 established financial standards and reporting requirements for all publicly traded U.S. corporations through the General Accepted Accounting Principles (GAAP). In 1973, the Securities Exchange Commission (SEC) assigned GAAP to the Financial Accounting Standards Board (FASB) which focuses on corporate stakeholders, investors, stockholders and creditors, as the principal users of financial statements, rather than corporate management.[3] GAAP establishes corporate financial statements as metrics in assessing a company's financial condition for a given period of time; i.e., annually, quarterly or monthly.

GAAP does not require firms to report free cash flow information. The GAAP provides the accounting rules used to develop the standard financial statements including the income statement, balance sheet, retained earnings and cash flow statements. However, many corporations do not include free cash flow on their cash flow statements. An *Accounting Horizon Journal* article published in December 2006 analyzed free cash flow information reported in Forms 10-K and 10-Q for 400 large companies, and concluded that companies including free cash flow in their statements signal positive financial performance to stakeholders and investors.[4]

Moreover, analyses of free cash flow in cash flow statements from leading U.S. corporations show its inclusion correlates with financial performance and a company's ability to produce "discretionary cash flow from operations."[5] Financial analysts and investors "...consider free cash flow to be cash that can be used for such purposes as debt reduction, dividends, stock buybacks, or acquisitions, while maintaining or even growing a company's productive

---

3   Christy, p 1.
4   Ajay Adhikari and Augustine Duru, "Voluntary Disclosure of Free Cash Flow Information", Accounting Horizons, Vol. 20, No. 4. (December 2006): pp. 311–332.
5   Charles W. Mulford and Kerianne Maloney, Corporate Reporting Practices for Free Cash Flow, DuPree Financial Analysis Lab, DuPree College of Management, Georgia Institute of Technology. (2004): p. 4.

capacity."[6] Stockholders and investors, and indeed, all corporate stakeholders, prefer companies "that produce plenty of free cash flow."[6] Free cash flow acts as a metric providing insight into reported earnings and corporate financial fundamentals.

For a small biotechnology enterprise, positive free cash flow can mean the difference between meeting strategic growth objectives, and lagging growth projections. A small enterprise with too much free cash flow may run the risk of being cash-rich but investment-poor in terms of building the corporate infrastructure. Corporate infrastructure includes investing sufficiently in plant equipment and human resources necessary for developing and sustaining growth. Conversely, insufficient free cash flow makes it difficult to cover necessary infrastructure investments.

However, this balance is the paradox of free cash flow; a company with a negative free cash flow position is not always an indication of a poorly run company. A company with positive free cash flow making large investments in its infrastructure could produce a short-term negative free cash flow position. The important point is that no matter the size of a company, the financial focus must be on establishing a positive free cash flow position. Sound financial management practices yielding strong earnings per share are a good indication of a well-managed company. However, the level of free cash flow signifies the financial viability of any sized company.

Financial management practices are essential for sustaining business operations, establishing solid growth and enhancing corporate value for the small biotechnology enterprise. A company initially funded by SBIR/STTR grants must look to create immediate cash flow through viable and readily available revenue streams such as through selling or licensing its technology. Divesting therapeutics and treatment practices can also create free cash flow. As a start-up transitions to the small growth company status, fundamental activities like understanding market demand, identifying competition, assessing customer needs and producing

---

6   Ben McClure, *Free Cash Flow: Free, But Not Always Easy*, Investopedia Advisor. (September 2003).

products or services meeting those needs should provide the revenue and the subsequent free cash flow in sustaining growth.

FINANCIAL STATEMENTS

The annual financial report is necessary in determining free cash flow since it contains the four primary financial statements: the income statement, the balance sheet, the statement of retained earnings, and the statement of cash flows. The income statement reports corporate revenue over a period of time. The income statement presents corporate revenues less operating expenses, yielding net income after interest and taxes are deducted (see Figure 1). Often referred to as a *profit-and-loss statement*, the income statement provides stockholders and investors an indication of how a corporation has performed over time. The two most important measures in determining free cash flow are net income (NI) and earnings before interest and taxes (EBIT).

Figure 1 provides an illustrative example of an income state-

Income Statement
Small Business Bio-Tech Company

Description	2006	2005	2004	2003
		[In Thousand U.S. Dollars]		
Total Revenue	10,284	8,430	6,853	5,664
Operating Expenses				
Costs excluding Depreciation	7,713	6,322	5,140	4,248
Depreciation	257	211	171	142
R&D	1,500	1,050	1,000	750
Total Operating Expenses	9,470	7,583	6,311	5,140
Operating Income or				
Earnings Before Interest & Taxes [EBIT]	814	847	542	524
Less Interest Expense	203	212	136	131
Earnings Before Taxes	610	635	407	393
Taxes [35%]	214	222	142	138
Net Income	397	413	264	256

**No Common or Preferred Stock Issued.

Figure 1: Income statement

ment for a hypothetical small biotechnology company over a four-year period between 2003 and 2006. EBIT and NI are the key income statement reporting metrics. In this illustration, the small biotechnology company has transitioned from start-up to what may be referred to as an emerging growth enterprise. Revenues increased from about $6 million in 2003 to more that $10 million in 2006. EBIT increased from $524,000 in 2003 (NI=$260,000) to over $800,000 in 2006 (NI=$400,000).

A potential investor presented with this income statement would notice NI increased a little more than fifty percent over the four year period. This stalwart number might arouse investment interest in the beginning. However, closer examination of the other three financial reports is also necessary. Given that investors look for additional reporting metrics of a company's financial strength, special attention is given to the level of free cash flow.

The balance sheet displays a company's assets, liabilities, and equity at a moment in time (see Figure 2). Assets are typically ordered in terms of their liquidity—the time it will take to transfer or convert them into cash. Assets are further categorized as current or long term. Current assets include all cash and equivalents, short-term investments, accounts receivable, and inventories. Fixed assets are long-term and consist of all property, plant and equipment. Intangible assets are also long-term assets as well as real estate held for investment.

Liabilities generally include accounts payable, notes payable and accruals. Long-term bonds are created for raising capital and typically carry a rating by either Moody's or Standard and Poor's. These thirty parties assess the risk and likelihood an issuer will default on interest or capital payments. Liabilities also include common stockholder's equity including any preferred or common stock issues. Preferred stock is a non-voting capital stock with a fixed dividend and follows corporate debt in the liquidation of assets. Common stock is the most typical class of stock issuance and comes with a voting right. However, it follows preferred stock in dividend allocations and liquidation rights. In contrast to preferred stock, common stock dividends can vary. Capital surplus, or the premium on capital stock, is the difference between the par value of common stock and an amount received for stock over and above the par value. It is displayed in the liability section of the balance sheet since it results from the issuance of common stock.

The statement of retained earnings is the third financial report for a company. This statement demonstrates management's

Balance Sheet
Small Business Bio-Tech Company

Description	2006	2005	2004	2003
	[In Thousand U.S. Dollars]			
Assets				
Cash and Equivalents	205	225	200	175
Short-Term Investments	-	-	-	-
Accounts Receivable	2,057	1,686	1,371	1,133
Inventories	2,571	2,107	1,713	1,416
Total Current Assets	4,833	4,018	3,284	2,724
Fixed Assets	3,919	3,706	3,438	3,317
Total Assets	13,585	11,743	10,006	8,765
Liabilities and Equity				
Liabilities				
Accounts Payable	3,599	2,950	2,399	1,982
Notes Payable	636	631	601	591
Accruals	5,235	4,400	3,741	3,183
Total Current Liabilities	9,471	7,982	6,740	5,756
Debt				
Long Term Bonds	-	-	-	-
Total Debt				
Equity				
Preferred Stock	-	-	-	-
Common Stock	100	100	100	100
Capital Surplus	900	900	900	900
Retained Earnings	3,114	2,761	2,265	2,009
Total Common Equity	4,114	3,761	3,265	3,009
Total Liabilities and Equity	13,585	11,743	10,006	8,765

Figure 2: Balance sheet

Retained Earnings
Small Business Bio-Tech Company

Description	2006	2005	2004	2003
	\[In Thousand U.S. Dollars\]			
RE Beginning Period (REBOP)	2,717	2,349	2,001	1,753
Add Net Income	397	413	264	256
Less Dividends	-	-	-	-
RE End Operating Period (REEOP)	3,114	2,761	2,265	2,009

Figure 3: Retained earnings

performance in handling company earnings by accounting for income and dividend payments (see Figure 3). Retained earnings represent claims against assets and indicates what is available, or retained by a company to expand business operations. The actual amount available for re-investment is contingent on dividend distribution during any one reporting period.

The fourth report is the cash flow statement of a company. It differs from the income statement and balance sheet in that it displays the flow of money from all operating activities for a given period. Cash flow is reported in terms of net cash from operating activities, financing activities, and investing activities (see Figure 4). A company's cash position reported by the balance sheet is affected by net income, non-cash adjustments, change in working capital, and fixed assets reported by the cash flow statement.

The statement of cash flow typically provides a company's cash position in three categories:
1. Net cash provided by operating activities.
2. Net cash provided by Investing activities.
3. Net cash provided by financing activities.

Net cash from operating activities includes net income and non-cash adjustments such as depreciation changes in working capital. Net cash provided by investing includes any investments made by a company in the sale or purchase of fixed assets. Net cash provided by financing activities includes cash raised by the sale of short-term investments like bonds and changes in notes payable. Of these three, net cash from operations provides the information financial managers look for in assessing trends in cash flow by examining the impact of changes in working capital on net cash.

Cash Flow Statement
Small Business Bio-Tech Company

Description	2006	2005	2004	2003
	[In Thousand U.S. Dollars]			
Operating Actitivites				
Net Income	397	413	264	256
Adjustments	-	-	-	-
Non-Cash Adjustments				
Depreciation [a]	257	211	171	142
Due to Change in Working Capital [b]				
Increase in Accounts Receivable	(371)	(315)	(238)	(173)
Increase in Inventories	(464)	(394)	(297)	(216)
Increase in Accounts Payable	649	552	416	302
Increase in Accurals	835	660	558	(130)
Net Cash Provided by Operating Activities	1,304	1,125	875	181
Long Term Investing Actitives				
Cash Used to Aquire Fixed Assets [c]	(470)	(526)	(292)	(309)
Net Cash Provided by Investing	(470)	(526)	(292)	(309)
Financing Activities				
Sale of Short Term Investments	-	-	-	-
Increase in Bonds Outstanding	-	-	-	-
Increase in Notes Payable	5	30	10	50
Net Cash Provided by Financing Activities	5	30	10	50
Summary				
Net Change in Cash	(20)	25	25	(50)
Cash At Beginning of Year	225	200	175	225
Cash At End of Year	205	225	200	175

Notes:

[a] Depreciation is a non-cash expense that is deducted when calculating net income. It must be added back to show the correct cash flow from operations.

[b] An increase in a current asset decreases cash. An increase in a current liability increases cash.

[c] The net increase in Fixed Assets is the net amount that is deducted for the year's depreciation expense. Depreciation expense should be added back to show the increase in gross fixed assets.

Figure 4: Cash flow statement

The summary provides the outcome of each reporting period's cash flow in terms of net change in cash. This is particularly important when a negative net change persisted over time necessitating a very close examination of the three categories of net cash shown in the statement of cash flows.

For example, Figure 4 illustrates the cash flows for a small biotechnology enterprise where the net change in cash at the beginning of years 2003 and 2006 are negative. Analysis shows that in each of the four years for which cash flows are reported, cash was used to acquire fixed assets. This is a negative entry on the cash flow statement. This company invested in fixed assets and a negative net change in cash at the beginning of a reporting period is not as troublesome as a significant increase in financing activities. An increase in financing activities could indicate increased risk and liability, disproportionately balanced against positive cash amounts at the end of the reporting period.

DETERMINING FREE CASH FLOW

The determination of free cash flow involves analysis of all four financial statements and the use of several key metrics. Operating assets and operating capital play an integral role in determining free cash flow.

Operating assets comprise cash, accounts receivable, inventories, and fixed assets. Furthermore, operating assets are pared down into two sub-categories: working capital, and fixed assets or plant and equipment. In addition to capital supplied by investors, accounts payable, accrued wages, and accrued taxes are a secondary source of capital. Accounts payable provide readily available cash for capital needs since the amount used to pay them can be deferred. This frees them up in the short-term. Wages earned and taxes incurred are another source of capital since they act as short-term loans against what is owed, and accrue before they are paid. Non-operating assets consist of assets not necessary for the operation of a company, such as marketable securities.

Free cash flow is derived from remaining capital or cash actually available for distribution to investors after the company has made all the investments in fixed assets and working capital necessary to sustain ongoing operations. The determination of free cash flow begins with determining the level of current assets (CA) and current liabilities (CL). Current assets, or operating working capital, include those assets used in the operation of a company and consist of cash, accounts receivable and inventories. Current liabilities are those liabilities that are to be paid within a year of being posted on the balance sheet, and consist of accounts payable and accruals. The difference between current assets and current liabilities yields the amount of actual capital needed for operations. This amount is referred to as *net operating working capital.*

(1.0) Net Operating Working Capital (NOWC)
(1.1) NOWC = CA - CL = (Cash+AR+Inv) - (AP + Accruals)

Net operating working capital is used in assessing a company's operational efficiency, where positive net working capital indicates a company's capability to meet short-term liabilities and operational strength. Negative net working capital suggests a company cannot meet short-term bills and may have management efficiency issues.

Figures 5 and 6 present the relevant financial information used in determining free cash flow. Totals may not add up to exact amounts due to rounding.

(1.2) CA = $4,833,000 and CL = $4,236,000, and
(1.3) $NOWC_{2006}$ = $4,833,000 - $4,236,000 = $597,000.

For 2005:
(1.4) CA = $4,018,000 and CL = $3,582,000, and
(1.5) $NOWC_{2005}$ = $4,018,000 - $3,582,000 = $437,000.

Figure 6 shows 2003 and 2004 also had positive NOWC, indicating NOWC for this fledging company increased steadily over the four year reporting period. Increasing NOWC suggests a well-

Small Business Bio-Tech Company

Description	2006	2005	2004	2003
	[In Thousand U.S. Dollars]			
Net Income [NI]	397	413	264	256
Depreciation	257	211	171	142
Fixed Assets	3,919	3,706	3,438	3,317
EBIT	814	847	542	524
Tax Rate	0.35	0.35	0.35	0.35
Cash	205	225	200	175
Account Receiveable	2,057	1,686	1,371	1,133
Inventory	2,571	2,107	1,713	1,416
Current Assets	4,833	4,018	3,284	2,724
Accounts Payable	3,599	2,950	2,399	1,982
Accruals	636	631	601	591
Current Liabilities	4,236	3,582	3,000	2,574

Figure 5: Determining Free Cash Flow

Free Cash Flow
Small Business Bio-Tech Company

Description	2006	2005	2004	2003
	[In Thousand U.S. Dollars]			
Operating Assets, Operation Capital and Operating Income				
Net Operating Working Capital [NOWC]	597	437	284	150
Total Operating Capital [TOC]	4,516	4,143	3,722	3,467
Net Operating Profit After Taxes [NOPAT]	529	550	352	341
Operating Cash Flow [OCF]	786	761	524	482
Net Investment in Working Capital [WC] in Yr	373	421	255	228
Gross Investment in Working Capital in Yr	630	632	426	370
Free Cash Flow [FCF] *Where*				
FCF = Operating Cash Flow - Gross Investment in Working Capital	156	129	98	113
Algebraically				
FCF = NOPAT - Net Investment in Working Capital	156	129	98	113

Figure 6: Free Cash Flow

managed operation representing a positive performance scenario for any future investor's consideration. A red flag would be raised suggesting possible management issues with the company or a possible risk investment scenario had NOWC been negative.

Total operating capital is the second metric needing calculation. This amount is derived from the sum of net operating working capital (NOWC) and fixed assets (FA) for a given reporting period. That is:

(2.0) Total Operating Capital (TOC)
(2.1) TOC = NOWC + Fixed Assets

Using the illustration of the small biotechnology company from Figures 5 and 6:
(2.2) TOC_{2006} = $597,000 + $3,919,000 = $4,516,000, and
(2.3) TOC_{2005} = $437,000 + $3,706,000 = $4,143,000

Between 2005 and 2006, the company increased its operating capital from $4,143,000 to $4,516,000, or by $373,000; where $213,000 is attributed to an increase in fixed assets and $160,000 is from an increase in NOWC.

The third metric in determining free cash flow is net operating profit before taxes, or NOPAT. NOPAT is an excellent measure of a company's operations and management effectiveness. That is:

(3.0) Net Operating Profit (Income) after Taxes (NOPAT)
(3.1) NOPAT = EBIT(1-Tax Rate)

From 3.1, NOPAT is a function of EBIT and measures operating income. NOPAT is important because it looks at what real profit would be if there were no financial assets or debt in a company. From the small company:
(3.2) $NOPAT_{2006}$ = $814,000(1.0-0.35) = $529,000, and
(3.3) $NOPAT_{2005}$ = $847,000(1.0-0.35) = $550,000.

Despite a slight decrease from 2005 to 2006, NOPAT remained

nearly constant. The significant change occurred between 2004 and 2005 where NOPAT increased from $352,000 to $550,000. This increase was mainly due to a 20% increase in revenues and a slight increase in R&D investment (See Figure 1).

The fourth metric in the determination of free cash flow is finding the level of operating cash flow or OCF. OCF is the sum of NOPAT and depreciation for a given period. That is:

(4.0) Operating Cash Flow (OCF)
(4.1) Operating Cash Flow (OCF) = NOPAT + Depreciation

From figures 5 and 6:
(4.2) OCF_{2006} = $529,000 + $257,000 = $786,000, and
(4.3) OCF_{2005} = $550,000 + $211,000 = $761,000.

The fifth metric or determinant in finding free cash flow is to calculate net investment in working capital for a given period.

(5.0) Net Investment in Working Capital (WC) in Yr_1
(5.1) Net Invest in WC = Change in TOC = TOC in Yr_2 - TOC in Yr_1

Plugging in data:
(5.2) Net Invest in $WC_{2006-2005}$ = [$4,516,000 - $4,143,000] = $373,000
(5.3) Net Invest in $WC_{2005-2004}$ = [$4,143,000 - $3,722,000] = $421,000

The sixth metric in calculating free cash flow is to determine the level of gross investment in working capital derived by adding depreciation to net investment in working capital.

(6.0) Gross Investment in Working Capital in Yr_i
(6.1) Gross Investment in WC in Yr_i = Net Invest in WC + Depreciation

Using the illustration of the small biotechnology company from Figures 5 and 6 then for illustrative purposes one more time:
(6.2) Gross Investment in WC_{2006} = $373,000 + $257,000 = $630,000, and

(6.3) Gross Investment in WC_{2005} = $421,000 + $211,000 = $632,000.

Now free cash flow can be determined with two methods using data garnered from the example.

(7.0) Free Cash Flow (FCF) where
(7.1) FCF = Operating Cash Flow - Gross Investment in Working Capital
(7.2) FCF_{2006} = $786,000 - $630,000 = $156,000, and
(7.3) FCF_{2005} = $761,000 - $632,000 = $129,000.

Algebraically:
(7.4) FCF = NOPAT - Net Investment in Working Capital
(7.5) FCF_{2006} = $529,000 - $373,000 = $156,000, and
(7.6) FCF_{2005} = $550,000 - $421,000 = $129,000.

Both of these methods work to determine free cash flow. They are the same since depreciation is actually added to both NOPAT and net investment. Using either method will yield the level of free cash flow. The metrics used in illustrating how free cash flow is determined, as well as the four financial statements comprising a corporate financial report, are necessary in determining free cash flow.

Including free cash flow on the statement of cash flow provides potential investors with critical investment information. Expanding free cash flow resulting from revenue growth and efficient operations notifies investors of increased earnings potential. It also showcases a strong investment opportunity. On the other hand, a consistently negative or declining free cash flow portends future trouble.

Poor free cash flow levels may cause the organization to increase its debt load to cover operation costs. This might demonstrate the company is not sufficiently liquid to remain a viable business entity. In this case, investors would likely not consider any level of investment. However, should this company have a particularly strong product or R&D program presenting significant technology, then the possibility of a merger or acquisition always exist.

Opportunities for outside investment are likely to remain low for the small biotechnology enterprise concentrating mainly on its technology. To be successful, the company must also establish a strong organization and financial structure promoting viable free cash flow levels. Creating strong free cash flow is essential in attracting outside investment to fund ongoing R&D and technology commercialization efforts.

Biotechnology Transfers and Models Facilitate Access to Biotechnological Inventions
Oleksandr Skorokhod

Oleksandr Skorokhod holds an MS in Biology and is a researcher in the in the Department of Cellular Signaling at the Institute of Molecular Biology and Genetics, National Academy of Science of Ukraine, Kiev. He has worked in biotechnology transfer and innovations at Birchbob since 2006, and he has served as a biotechnology expert in Belgian-Ukrainian company Intellectual technology since 2007. Hobbies include karate do, chess, and mountain hiking. Olexandr can be contacted at fiwinner@ukr.net.

The ownership and exploitation of intellectual property rights (IPRs) is the key factor in determining the success of any technological innovation introduced in the market that provides the means for technological progress. The efficient management of IPRs is therefore crucial in providing incentives for continuing technological innovations.

Patents are widely seen as the lifeblood of biotechnology companies, who are dependant on attracting venture capital and angel investment for further research and development, and on being able to license technological developments to downstream product developers. However, a great number of biotechnological patents (in genetics for instance) are particularly controversial because they lie at the interface between discovery and invention and signal a move away from patenting end products toward patenting basic scientific information.[1]

In order to overcome the difficulties associated with biotechnology patents, several national, regional and international organizations together with scientists, the pharmaceutical industry and academics are debating alternative licensing models, which may help in technology access and transfer. These alternative models aim to allow effective use of diagnostic testing services, essential in the light of public health, and to enable further research on related technologies. The two mechanisms attracting most interest are clearing houses and patent pools.

CLEARING HOUSES

The term "clearing house" is derived from banking institutions and refers to the mechanism by which cheques and bills are exchanged among member banks to transfer only the net balances in cash. Nowadays the concept has acquired a broader meaning referring to any mechanism by which providers and users of goods, services and/or information are matched.[2]

At least five fundamentally different clearing-house models exist: the information clearing house, the technology-exchange clearing-house, standardized licenses clearing house, open-source clearing houses, and royalty collection clearing house. Basically, all clearing houses perform one or more of the following functions: facilitating the search for technology available for licensing

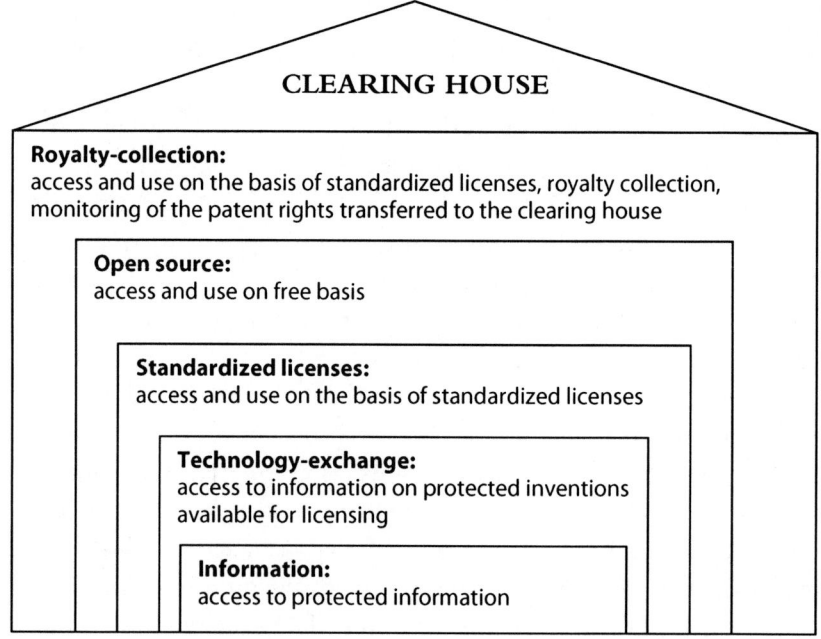

Fig.1. Clearing-house models

or free use; smoothing the progress of negotiations; and monitoring or enforcing negotiated agreements.[3]

INFORMATION CLEARING HOUSE

The information clearing house provides a mechanism for exchanging technical information and/or information that is related to the IP status of that information. Information mechanisms are relatively easy to set up but require constant maintenance and updating.[2,4]

Science clearing houses include freely accessible (European Patent Office (EPO)) or fee-based (Delphion, STN International, Micropatent) patent search sites. In addition, there are specific search platforms for biotechnological patents, such as Patent Lens, which is a text-searchable database of USA, European, Australian and international agricultural and life-science patents, and is complemented by advisory and educational services.[5]

TECHNOLOGY-EXCHANGE CLEARING HOUSE

The technology-exchange clearing house is inspired by the internet-based business-to-business (B2B) model.[5] This model provides an information service that lists the available technologies allowing technology owners and/or buyers to initiate negotiations for a license. Additionally, it may provide more comprehensive mediating and managing services.[4,6] An example of a global technology-exchange model is BirchBob, which is an internet-based platform that brings together offers and demands for innovations, and provides services dedicated to finding and facilitating contacts between technology holders and technology seekers. More than 40,000 innovations from about 2,000 organizations worldwide are currently searchable on BirchBob by investors, entrepreneurs and scientists who are looking for new business or scientific opportunities.

Specific health-care technology platforms include Pharmalicensing and TechEx, which provide online support for partnering and licensing in the biopharmaceutical and biomedical industry. Specific biotechnology platforms include the Public

Intellectual Property Resource for Agriculture (PIPRA)—a collaboration among universities, foundations and non-profit research institutions aiming to make agricultural technologies more easily available.[5]

STANDARDIZED LICENSES CLEARING HOUSE

The standardized licenses clearing house provides access to, and standardized licenses for, the use of protected inventions. Science Commons is a good example of this model. The Science Commons Licensing Project aims to develop a standard open framework for managing transfer of research materials including cell lines, animal models, DNA constructs and screening assays.[6] Its sister organization, Creative Commons, has successfully facilitated the use of copyrighted material (such as video, photos, books including scientific). Science Commons is more advanced than the Creative Commons, as it does not merely link offers to demands, but also provides standardized licenses worldwide.[5]

OPEN SOURCE CLEARING HOUSE

A fourth and unique model is the open source clearing house fosters the free exchange of technology. Open-source licensing is essentially the licensing of inventions without patent protection—the only requirement is that any licensee must agree to also make available to others any improvements in the invention or technology. Perhaps there is a need for a biotechnology clearing house, whereby anyone can post biotechnology inventions that are not IP protected.

The Human Genome Project (HGP) provides a good illustration of the way this open access approach can be employed in genomics. The HGP commenced in 1990, and from the outset was a collaborative venture, both between institutions and between countries. The goals of the HGP were to map all of the genes and to systematically sequence the genetic code for the entire human genome. In 1996 HGP participants agreed in the Bermuda Declaration that primary genomic sequences should remain in the public domain and should be rapidly released.[7] GenBank is the publicly accessible repository of the sequence information produced by the HGP. There are a number of advantages to be gained by putting this information in the public domain: first, it reinforces the norm of open science; secondly, it devalues competing proprietary sequence databases; and thirdly, it effectively excludes the patenting option until some additional steps are taken, for example ascribing function to a particular gene sequence. It should be noted that the patentability of raw sequence data is questionable in any case, because it does not satisfy the requirement for industrial applicability or utility.

In addition to the HGP, there are a number of other international collaborative sequencing ventures. Notable examples include the SNP Consortium and the HapMap Project. Both also make sequence information available in publicly accessible databases.[8] For instance, the goal of the this organization is to identify and collect SNPs, and create and make publicly available a map of all catalogued SNPs of the human genome without proprietary rights being retained by the members of the consortium. This method allows further drug discovery.

ROYALTY-COLLECTION CLEARING HOUSE

The royalty-collection clearing house comprises major aspects of the technology-exchange scheme, with the addition of managing collection of license fees from users on behalf of the patent holder in return for the use of certain technologies or services.[9] The patent holder is reimbursed by the clearing house pursuant to a set allocation formula. At present, there are no working examples in the field of patents and genetic inventions, although there are a number examples of such clearing houses in other sectors.[10-14] An attempt to design a royalty collection clearing-house model in the life sciences—the Global Bio-Collecting Society (GBS)—did not materialize, probably because no consensus could be reached among the stakeholders and because the necessary political support was missing.[15] The GBS was designed to be an efficient, fair and equitable model for the exchange of indigenous knowledge between knowledge holders (indigenous groups) and knowledge users (the life-science industry) in the commerce of biodiversity.[6]

STRENGTHS AND WEAKNESS

All clearing house models have, of course, some advantages as well as disadvantages, compared to each other.

Information clearing houses are relatively easy to set up but rather complicated to maintain and update. Information clearing houses also provide access to a country's biotechnology, just as training clearing houses offer training for biotechnology technicians, and industry links, updates, news, and job markets.[2]

The technology-exchange clearing-house model is generally cheap to maintain and has low operating costs. However, it maybe difficult to aggregate the critical mass of genetic patents needed to turn platforms of this type into useful tools. At present, most technology-exchange platforms cover only a small proportion of the market and a low density of patents, and one has to search various web sites (sometimes paying considerable registration fees) to find desired patents. Moreover, this model might only be suitable for technologies that can be easily defined and valued. Therefore, it might be a limited model for general-purpose research methods, such as PCR, and for patents that protect specific and well-defined improvements to familiar downstream products or processes.[2,4]

A standardized licenses clearing house is a model capable of lessening the legal and technical costs of sharing and reusing scientific work. But standardization in the biotechnological sector is problematic.

Open-source clearing houses may be a readily available model for sharing and exchanging unpatented technology. However, most genetic inventions are the outcome of long-lasting research requiring high levels of investment. Both private enterprises and universities wish to recover those investments and consequently apply for patent protection. Therefore, the scope of application for this model might be limited to the area of genetic inventions, at least in the near future.[2]

A royalty-collection clearing house is be more complicated to set up in comparison to the other clearing-house models; however once established, it can facilitate the collection of royalties. Although the concerns of the authorities overseeing free competition might vary according to the actual legal structure chosen for the clearing house (for example, a private entity that comprises patent holders as its members, or a neutral, independent, public clearing institution), one should always be aware of potential anti-competitive effects.[6] Furthermore, this type of clearing house is only useful if there is a recurring need to transact in the included patents, and if many patent holders or an entire branch of industry would participate.

PATENT POOLS

One of the biggest public concerns voiced against the United States Patent and Trademark Office (USPTO) for its practice of granting patents for inventions in biotechnology, particularly in genomics, is the difficulty of accessing patented inventions for basic biological research and R&D.[2] One solution to this constraint is to form patent pools, a mechanism successfully implemented by other industries.

A patent pool is an agreement between two or more patent owners to license one or more of their patents to one another, or to license them as a package to third parties who are willing to pay the royalties associated with the license. Patent pools can be a useful model to gain access to patented technology when access and use of technology are hindered by the existence of multiple patents held by multiple patent owners.[16]

Licenses are provided to the licensee, either directly by the patentee, or indirectly through a new entity specifically set up for the administration of the pool.[17] Therefore, a patent pool is formed by the patent holders, who act as shareholders of the pool, and as financiers of the licensing entity. Consequently, patent holders retain authority over licensing conditions.[18]

The establishment of patent pools in genetics was suggested by the Organisation for Economic Co-operation and Development (OECD).[5] The OECD considers the patent-pool concept to be interesting for biotechnology, but calls for further study. In February

2006 the OECD Council approved *Guidelines for the Licensing of Genetic Inventions*.[3] In summary, these guidelines aim to foster innovation by achieving a balance between return on investment on one hand, and dissemination of information and access to healthcare products on the other. According to the guidelines, best practice generally requires broad licensing of genetic inventions for research and investigation, and licensing for health applications on such terms and conditions that ensure widest public access to healthcare products and services. As a general rule, the OECD recommends inventions be non-exclusively licensed. Although in some limited circumstances, exclusive licensing may be appropriate, provided sufficient safeguards are in place to ensure the invention is sufficiently exploited.

But the OECD fears biotechnological companies rely too heavily on their intellectual property (IP) and foster what has been called a "bunker mentality." This mentality causes difficulties in the process of creating a pool.[2]

Nevertheless, there are already some patent pools in genetics. A first instructive genetic patent pool, which gained wide attention, is the Golden Rice Pool[5]. While working at the Swiss Federal Institute of Technology, Ingo Potrykus succeeded in genetically enriching rice grains with ß-carotene[19], and wanted to transfer the golden rice materials to developing countries for further breeding by introducing the trait into local varieties that are consumed in selected countries. Six key patent holders struck an agreement allowing Potrykus to grant licenses, free of charge, to developing countries, with the right to sub-license.[20] This agreement is an example of how private and public organizations, in a combined effort, dealt with the surrounding patents to create a non-profit humanitarian (and therefore probably atypical) patent pool in the form of a single licensing authority.[21]

Another genetic pool, supported by the World Health Organization (WHO), is under way — the SARS (Severe Acute Respiratory Syndrome) corona virus pool.[5] Relevant patent holders have been identified and agreement has officially been gained by signing a letter of intent (J. Simon, personal communication).

The SARS pool highlights the opportunities offered by the patent-pool concept for biomedical genetic inventions.

Patent pools comprising sequence data for genetic testing purposes are also worth investigating. Most susceptible to the patent-pool concept are cases involving a disease caused by various mutations in one gene, or by one or more mutations in any one of several possible genes. These cases are more likely to give rise to patent thickets.[22] However, it remains to be seen whether a gene patent pool covering only one disease syndrome will reach a fair balance between the costs of creating a pool and adequate revenue, and whether small pools prove to be viable.

As well as providing a possible solution to the problem of patent thickets, the creation of a patent pool might also stimulate funding for research and development, benefiting all partners in the pool.[23] As has been demonstrated in the electronics and telecommunications sector, the main incentive to establishing a patent pool is the generation of an internationally accepted technical standard. It has been claimed that such a standard is missing in genetics.[24] However, in the context of genetic testing, standards could be defined by establishing a set of mutations that are recognized by the international scientific community, or by reflecting national or international best practice guidelines relating to genetic testing for a particular disease.[25]

The following types of patent pools could be established today[2]:

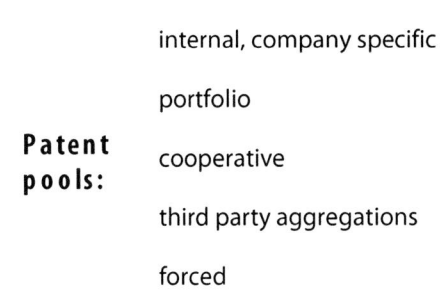

Figure 2: Types of patent pools

INTERNAL, COMPANY SPECIFIC

For example, DuPont combines technologies through internal development or Syngenta complements its internal portfolio with outside technology through licensing and M&As. The critical challenge is keeping internal innovation ongoing and tightly managed.

PORTFOLIO POOLING

Internal technology is supplemented with third party technologies (e.g., Microsoft and others). The critical challenge is having a dynamic team handling in-licensing and aligning strategies closely with the overall corporate strategy.

COOPERATIVE POOLING

Companies agree to combine their technologies and allow them to be managed by a separate entity, typically for standard-setting purposes. The critical challenge is to avoid anti-trust issues.

THIRD PARTY AGGREGATIONS

This strategy is practiced by BTG Ltd. The critical challenge is working around anti-stacking provisions that are very common in biotechnology licenses.

FORCED POOLING

Rarely-enforced compulsory licensing is one example. Another example is the pooling forced by the US government shortly after radio was invented.

This list is not exhaustive. New types of patent pools, based on other concepts, could be established.

BENEFITS AND RISKS

Patent pools might have significant benefits: elimination of stacking licenses; reduction of licensing transaction costs with the introduction of a system of 'one-stop licensing' for non-member licensees; decrease in patent litigation; and institutionalized exchange of technical information not covered by patents. This exchange would operate through a mechanism for sharing technical information relating to the patented technology, which would otherwise be kept as a trade secret.[18,26] Furthermore, patent pools offer an interesting instrument for government policy: it is better to encourage companies to establish patent pools than to force them into a compulsory licensing scheme.[18] Such a suggestion seems to ignore the fact the main prerequisite for establishing patent pools is the voluntary participation of all patent holders, whereas the compulsory licensing mechanism is the last-resort instrument for patent holders who do not voluntarily wish to enter into (reasonable) licensing negotiations.[5]

For developing countries, patent pools may be even more important because companies can easily obtain the licenses required to practice a particular technology, which reduce transaction costs and facilitate the rapid deployment of new applications.

Patent pools might carry some risks: they can shield invalid patents and entail the risk of inequitable remunerations, although expert valuation could settle disagreements on the value of the patents.[17] Additionally, patent pools can shield cartel cooperativity and, subsequently, have anti-competitive effects.[27,28]

SUMMARY

Today there are two main mechanisms which facilitate access to innovation in biotechnology: clearing houses and patent pools. Each model has several strengths and weaknesses, and can improve access to inventions in the global technology market. It is unlikely any one mechanism will provide an unambiguously perfect solution; more likely, the use of different models will benefit the biotechnological industry. Clearing house models and patent pools only recently emerged, and there will need to be a period of experimentation before they can be fully evaluated. Future changes in technology protection needs will require making changes in the existing models as well as the creation of novel, more appropriate models.

REFERENCES

1. Eisenberg, R. (2002) How Can You Patent Genes? The American Journal of Bioethics 2:3.
2. Krattiger, A. F. (2004) Financing the bioindustry and facilitating biotechnology transfer. IP Strategy Today 8:1–45.
3. OECD, Guidelines for the Licensing of Genetic Inventions (2006) OECD, Paris.
4. Graff, G. and Zilberman, D. (2001) Towards an intellectual property clearinghouse for ag-biotechnology. IP Technol. Today 3:1–12.
5. Van Overwalle, G., van Zimmeren, E., Verbeure, B., Matthijs, G. (2006) Models for facilitating access to patents on genetic inventions. Nature Reviews Genetics 7:143-154.
6. van Zimmeren, E., Verbeure, B., et al (2006) A clearing house for diagnostic testing: the solution to ensure access to and use of patent genetic inventions? Bulletin of World Health Organization 84:352-359.
7. Bentley, D. R. (1996) Genomic Sequence Information Should Be Released Immediately and Freely in the Public Domain Science 274:533.
8. Chokshi, D. A. and Kwiatowski, D. P. (2005) Ethical Challenges of Genomic Epidemiology in Developing Countries 1/1 Genomics, Society and Policy 1.
9. Merges, R. P. (1996) Contracting into liability rules: intellectual property rights and collective rights organizations. Calif. Law Rev. 84:1293–1386.
10. Australian Law Reform Commission. Gene patenting and human health, discussion paper 68. Australian Law Reform Commission web site [online], <http://www.alrc.gov.au> (2004).
11. Gold, E. R. (2002) Biotechnology patents: strategies for meeting economic and ethical concerns. Nature Genet. 30:359.
12. Nuffield Council on Bioethics. The Ethics of Patenting DNA: A Discussion Paper (2002).
13. Organisation for Economic Co-operation and Development. Draft guidelines for the licensing of genetic inventions (point 39 and 46). Organisation for Economic Co-operation and Development web site [online], <http://www.oecd.org/sti/biotechnology/licensing> (2005).
14. HUGO Intellectual Property Committee. Statement on the scope of gene patents, research exemption and licensing of patented gene sequences for diagnostics. The Human Genome web site [online], <http://hugo.hgu.mrc.ac.uk/committee_ip.htm> (2003).
15. Drahos, P. (2000) Indigenous knowledge, intellectual property and biopiracy: is a global bio-collecting society the answer? Eur. Intellect. Prop. Rev. 20:245–250.
16. Clark, J. et al. Patent pools: a solution to the problem of access in biotechnology patents? United States Patent and Trademark Office White

Paper [online], <http://www.uspto.gov/web/offices/pac/dapp/opla/patentpool.pdf> (2000).
17. Klein, J. I. Business review letter to Gerrard R. Beeney, [online], <http://www.usDoJ.gov/atr/public/busreview/1170.htm> (1997).
18. Merges, R. P. in Expanding the Boundaries of Intellectual Property (eds Dreyfuss, R., Zimmerman, D. L. & First, H.) 123–166 (Oxford Univ. Press, 2001).
19. Beyer P. et al. (2002) Golden Rice: introducing the ß-carotene biosynthesis pathway into rice endosperm by genetic engineering to defeat vitamin A deficiency. J. Nutr. 132:S506–S510.
20. Zeneca (now Syngenta) media release. Golden Rice collaboration brings health benefits nearer. Syngenta In The News web site [online], <http://www.syngenta.com/en/media/article.aspx?pr=051600&Lang=en> (2000).
21. Parish, R. and Jargosh, R. (2003) Using the industry model to create physical science patent pools among academic institutions. J. Assoc. Univ. Technol. Managers 15:65–79.
22. Scherer, F. M. (2002) The economics of human gene patents. Acad. Med. 77:1348–1367.
23. Klein, J. I. Business review letter to Gerrard R. Beeney, [online], <http://www.usDoJ.gov/atr/public/busreview/2121.htm> (1998).
24. James, C. A. Business review letter to Ky P. Ewing, [online], <http://www.usdoj.gov/atr/public/busreview/200455.pdf> (2002).
25. Ebersole, T. J., Guthrie, M. C. & Goldstein, J. A. (2005) Patent pools and standard setting in diagnostic genetics. Nature Biotechnol. 23:937–938.
26. Shapiro, C. in Innovation Policy and the Economy Vol. 1 (eds Jaffe, E., Lerner, J. & Stern, S.) 119–150 (MIT Press, 2001).
27. James, C. A. Business review letter to Ky P. Ewing, [online], <http://www.usdoj.gov/atr/public/busreview/200455.pdf> (2002).
28. Carlson, S. C. (1999) Patent pools and the antitrust dilemma. Yale J. Regul. 16:359–399.

Maximizing the Strategic Impact of Health- and Pharmacoeconomics in Biotechnology Companies

Ulf Staginnus and Stephen Russell

Ulf Staginnus is an economist with more than 10 years of international experience in health economics, pricing, and reimbursement strategies as well as competitive intelligence within the biotechnology and pharmaceutical industry (Bristol-Myers Squibb, Pfizer Inc., Baxter Intl. Inc., GSK Biologicals). He also founded and runs his own consultancy serving biotech and pharma clients in supporting product pricing, valuations and market access strategies. Ulf has been profiled in Marquis Who's Who in America, Healthcare & Emerging Leaders for his achievements in the area of health- and pharmacoeconomics. He may be contacted at *ustaginnus@hotmail.com* or *www.healtheconomicsblog.com*.

Stephen Russell is a consultant for the European pharmaceutical and medical device markets in the areas of health economic strategy and value demonstration. His clients range from large established multinationals to start-ups. He has also worked as a consultant for private equity firms valuing potential acquisitions in the health care area. Stephen may be contacted at *stephen_russell@telefonica.net*.

Biotechnology medicine development is growing and will continue to drive both start up and established pharmaceutical company pipelines for years to come. Many of these new innovative treatments have improved patient outcomes in diseases such as cancer, rheumatoid arthritis and infectious diseases. The high complexity associated with treatment development, as well as formulary management and administration requirements, necessitates high costs. Runaway prices and rising budgetary constraints concern healthcare administrators around the globe.

Internal outcomes research and health economics capabilities have long been established in big pharma. Biotechnology companies, especially small and start up companies with their drug-discovery-centred focus, will need to catch up in order to meet tougher market access environment requirements. In this chapter we highlight the commercial importance of incorporating health economics into broader business strategies in a systematic fashion to meet the needs of a rapidly changing external payor environment.

US BIOTECH – THE NEED TO DEMONSTRATE THE VALUE OF ITS BIOPHARMACEUTICALS

There has always been discussion of economic value proposition concepts for payors. A payor is a person who pays bills, or reimburses for new technology in a health plan. In reality, at least from a health economic perspective, it seems not much has happened concerning small companies. The Brucker Group (BGI) authored a 2006 report regarding changing attitudes in healthcare costs. BGI encourages US biotech companies to understand the pressures and to start providing their main customers (payors such as managed care organizations, pharmacy benefit managers, Medicare, and Medicaid) with health economic data to assess the value of their products. Providing pharmacoeconomic data is good business, and companies need to be prepared to demonstrate value from the outset.

Initially, so called 'value-based' analysis performed by payors was limited to *me-too* drugs. Finding cost comparison and efficacy in a cost-minimization fashion was relatively straightforward. Because biotech products were traditionally targeted towards diseases such as cancer within specific, limited populations, coverage by third-party payors had been practically guaranteed due to the low impact on the overall health care budget. With the emergence of new biotech products aimed at larger patient populations with chronic conditions, payors will be looking more at a drug's overall health economic value in deciding whether or not to reimburse. Therefore, pharmacoeconomics and health technology assessments of biotechnology drugs will become increasingly important as a management tool for health plans.[1]

Small pharmaceutical and biotech companies have been particularly slow to adapt to payors' cost concerns, according to BGI.

---

1   For simplification, the terms "healtheconomics", "pharmacoeconomics" and "health technology assessment" will be used interchangeably.

Managed care organisations (e.g. WellPoint) are now using their buying power as leverage in asking drug companies to voluntarily submit a data dossier measuring treatment outcome. The data from clinical trials and postmarketing studies is helpful in determining the drug's pharmacoeconomic value. Almost all payors surveyed by BGI argue biotech companies are not yet ready to fulfil this requirement.

THE EMERGING HTA AGENCIES LANDSCAPE IN EUROPE

In the past only Canada, Australia, and the UK have established HTA (health technology assessment) agencies. These agency formally assess the economic value of new medical interventions and link their decisions to reimbursement coverage. However, in the last couple of years, there has been a real 'boom' of HTA assessment agencies emerging in the European Union. Their emergence is motivated by exploding health care budgets for drugs and the simultaneous desire for cost-containment:

A number of countries and states are now requiring explicit economic evaluations of new techniques and there to be undertaken as part of the approval and licensing process or access provisions that influence use and pri For example in 2007 :

Figure 1: Countries that require economic evaluations for either price negotiations, reimbursement decision making and/or formulary listing, 2007
Author's elaboration

Considering the current environment, a biotechnology executive will need to have an appreciation of the areas where health economics plays a role for his or her organization. In the past, the commercial potential of a drug was based primarily on its effectiveness against a disease or health condition. Today, the link between that effectiveness and its cost is what matters. We will briefly describe the main analytical concepts of health economics before getting into a more detailed description of using health economics to maximize a drug's commercial potential.

AN INTRODUCTORY PRIMER TO HEALTH ECONOMICS

A full economic evaluation deals not only with costs but also with the consequences (health benefits) of the intervention studied. Costs are always expressed in monetary units (e.g. dollars). The distinguishing feature among economic evaluation techniques is the way in which health benefits are expressed: in monetary units, in natural units (cases avoided or life-years saved), or in quality adjusted life years (QALYs).

TYPE OF ECONOMIC ANALYSIS FOR NEW PRODUCTS

"Cost-benefit analysis" is sometimes used loosely as a general term covering all types of economic evaluation. Among health economists, the term is usually restricted to those forms of evaluation in which the health benefits (cases avoided or life-years saved) are expressed as a monetary value. Not surprisingly, the main controversy surrounding cost-benefit analysis is how monetary values are attached to benefits such as health outcomes. This is generally achieved using a "willingness-to-pay" approach. These values are calculated by eliciting the value individuals might place on reduced morbidity or on an intervention's impact on quality of life. Cost-benefit analysis forces an explicit decision between costs and benefits by measuring both in the same units. Some cost-benefit or net-present valuations have been conducted, especially in the area of vaccines. These are not true cost-benefit studies in the sense of economic theory (e.g. benefits have been measured in an accounting fashion rather than based on opportunity cost through a willingness to pay concept). Instead, they are based on a subtraction of direct and indirect costs.

COST-EFFECTIVENESS ANALYSIS

In cost-effectiveness analysis, the problem of placing a monetary value on life or health is avoided. Incremental costs are compared to incremental health benefits, as measured in natural units, such as cases avoided, disability-years averted, or life-years saved. The effectiveness of different procedures is then expressed in terms of cost per unit outcome. A particular type of cost-effectiveness analysis is cost-utility analysis, in which outcomes are measured in terms of QALYs gained. QALYs combine changes in quantity and quality of life into a single composite measure independent of the program or disease being assessed. This approach makes it possible (at least theoretically) to compare diseases in so called QALY league tables. The quality-adjustment factors (or utilities) are weights ranging from 0 to 1 (1, optimal health; 0, health state judged to be equivalent to death). These factors should reflect aggregated preferences of individuals for the outcomes and have been measured directly among patients or the general public.

Cost-effectiveness analysis using QALYs is the most common analyses. QALYs are the preferred units for HTA bodies in Europe such as NICE (National Institute of Clinical Excellence). The use of QALYs has been recommended by the Panel on Cost-effectiveness in Health and Medicine (Harvard University) in the US.

COST-MINIMIZATION ANALYSIS

A cost-minimization analysis is a type of cost comparison study involving two or more treatments considered to be comparably effective in terms of clinical and quality-of-life outcomes. Only economic cost is a differentiating factor. When two or more interventions result in identical outcomes, a cost minimization analysis is a suitable tool for deriving the cost associated with each outcome. Because the outcomes of two different drugs are rarely, if ever, equal, this type of study is applicable and useful for evaluating different dosage forms of the same drug. It can also be used for evaluating generically equivalent drugs for which outcomes have been demonstrated to be equivalent. A preference is then made between two or more alternatives based on which one costs less.

Budget impact analysis

A product may not be cost-saving even if it is cost-effective (resulting in a cost per QALY ratio below commonly accepted thresholds for cost-effectiveness). Budget impact analysis is therefore an essential part of a comprehensive economic assessment of a health care technology and is increasingly required, along with cost-effectiveness analysis, prior to formulary approval or reimbursement. The financial consequences of new drug adoption and diffusion on the health care system can be estimated with this analysis. A budget impact model predicts how a change in the drug and therapy mix used to treat a particular health condition will impact spending on that condition. It can be used for budget planning, forecasting and for computing the impact of health technology changes on premiums in health insurance schemes. In essence, it is used to answer the question: can we afford this? Users of budget impact analyses include those who manage and plan for health care budgets. These include administrators of national or regional health care programs, administrators of private insurance plans, administrators of health care delivery organizations and employers who pay for employee health benefits. Each has need for financial impact information concerning alternative health care interventions, yet each has different and specific evidentiary requirements for data, methods and reporting.

The choice of perspective

Defining cost components depends on perspective. Who is using the product and who benefits? This important part of economic evaluation determines which costs should be included in the analysis and how they should be valued. Whether a drug is cost-effective may depend on one's perspective. The patient, society in general, and a third-party payor may reach different conclusions about specific costs. For example, direct costs like hospitalization are covered by payors but indirect costs due to work loss are born by society or by employers. In a comprehensive societal perspective, all costs and benefits should be identified, regardless of who incurs the costs and who receives the benefits.

Certain diseases like rheumatoid arthritis are equally burdening. The indirect cost due to productivity loss, work loss, or caregiver time is considered from the perspective of society as a whole.

Models and their application

Most economic evaluations, especially in the early phases of product development, are based on analytic models rather than on real observational data. The main advantage of using these models is their flexible and timely framework for analysis. Models have their limitations, however. Pieces of information from different studies and populations are combined and used in a single model. This introduces higher risk of uncertainties, or even bias. Expert estimates may have to be incorporated into the analysis due to a lack of real data. Because of the high complexity of many models, (i.e. is the model accurately replicating the natural history of a disease and/or potential real life effects?) validations are than also more difficult to conduct. The data generated from a model must be interpreted within the limitations of the model and its assumptions.

Management of study uncertainties

Economic evaluations require the analyst to combine information about the epidemiology of the disease, the probabilities of clinical consequences, the clinical effectiveness of the drug, and incurred costs. Incurred costs may include treatment of the disease and its consequences, administration, and treatment of potential adverse effects. Sensitivity analysis is the main method by which health economists test for uncertainty. Sensitivity involves changing the value of variables that are known to be uncertain or to change over time. A plausible range for variation can be determined by review of the literature or by consulting experts.

Another approach uses scenario analysis. Typically, the scenarios will include a base-case (best guess) scenario, the most optimistic (best-case) scenario, and the most pessimistic (worst-case) scenario.

Finally, another approach is to undertake a threshold analysis. In threshold analysis, the critical value(s) of a parameter or parameters central to the decision are identified. The analyst assesses which combination of parameter estimates could exceed the threshold and make the program unacceptable. Multivariate sensitivity (varying model parameters using likelihood distributions) analysis and patient level simulations (simulating the potential outcome on a patient rather than at cohort level) are other techniques aiding the calculation of uncertainty. This results in a more precise estimate of the uncertainty intervals.

WHEN CAN A NEW MEDICAL INTERVENTION BE CONSIDERED COST-EFFECTIVE?

There is no single, valid criterion for evaluating the cost-effectiveness ratio (the ratio between the difference of effects divided by the difference in cost between a new and standard alternative). Ratios needn't be below any level for adoption. A threshold value of $50,000 per QALY gained is often quoted in the US medical literature. Other thresholds utilized are £30,000 per QALY by the UK NICE, or the unofficial threshold of €30,000 per QALY for a cost-effective health technology used in other EU countries such as Spain.

Another proposed threshold has been made by the WHO. Gross domestic product (GDP) is a readily available indicator and can be used to derive three categories of cost-effectiveness: Highly cost-effective (less than GDP per capita); Cost-effective (between one and three times GDP per capita); and Not cost-effective (more than three times GDP per capita).

HEALTH- AND PHARMACOECONOMICS—WHAT A BIOTECH EXECUTIVE NEEDS TO KNOW ABOUT IT

HEALTH ECONOMICS & DRUG PRICING

Pharmacoeconomic information is not only important to external audiences but also has internal audience. Internal decision-making, for example, determines the relation between price and potential cost-effectiveness of a new development compound. Biotech companies have considered development risk only. Now, marketing (market access) risk is based on the so called fourth hurdle (being rejected formulary access and/or reimbursement on cost-effectiveness or budgetary impact grounds). Rejection is due to a more careful evaluation of new products for formulary access in managed care organizations in the US and the introduction of health technology assessment agencies in Europe. Including this market access dimension is a key factor to consider in developing a pricing strategy. Moreover, payors across the world have become powerful players in the market. Biotech companies need to better understand their needs and make a compelling case for health economics. These considerations need to move to the front-end of the commercial strategy. Strategic placement of economics will gain momentum with countries becoming serious about economic evaluations as a part of their pricing and reimbursement decision making.

Biotech product prices are often based on estimates and "gut feelings." Sometimes competitor's prices are taken into consideration. Prices do not necessarily reflect the product's perceived value or the payor's value. Correct pricing is a necessary component of due diligence. The question is how to go about it?

Payors will only grant reimbursement if a product's projected cost-effectiveness will fall into a more or less defined threshold for willingness to pay (WTP). This is about £30,000 per quality adjusted life-year (QALY) in the UK, less than \$50,000 per QALY in the US, and about €30,000 per QALY in other EU countries. A new biotech compound can be described as a function of its cost-effectiveness. Under this assumption a drug can be successfully marketed when its:

$$CE \leq \lambda$$

Where:
CE = Cost-effectiveness
λ = Threshold (Willingness to pay) for cost-effectiveness

The envisioned price is a key variable in the nominator of the cost-effectiveness ratio when comparing one or more alternative treatments. In light of cost-containment in all healthcare systems, value-based pricing strategies are most likely to be preferred for new biopharmaceuticals whereby:

$$V = R \pm D$$

Where:
V = Value (Price)
R = price of reference product or best alternative
D = net value of perceived differentiation vs. standard therapy

The expected return on investment defines the lower launch price band (minimum price) while D defines the potential incremental value of the product to stake holders in the healthcare system. In recent years, better economic modelling methods have evolved to more accurately define D, or at least indicate likelihood of achieving a certain D. D values facilitate early analysis (e.g. phase II trials) describing the expected price-effectiveness relation of new technologies compared to λ.

So what does this all mean to a biotech executive concerned with drug pricing? Executives must understand the potential trade-off between a price that is meeting internal return on investment criteria, and a price that will be considered cost-effective by payors. Relying on gut-feelings when making initial pricing assumptions may leave money on the table, meaning the product could have been priced higher and would still be cost-effective. The company may price the product too high by creating unrealistic expectations. Company investor expectations will face harsh reality when they meet market (payor) reality.

A pricing assumption based on systematic health economics analyses is important for the launch sequence of companies targeting different indications with the same drug. The product may be much more cost-effective (and hence more valuable) in one indication than is characterized. The product may be cost-effective by focusing on unmet medical need and/or more serious consequences. The company would ideally want to launch first in the indication where the product could command a higher price. This concept becomes even more pronounced for targeted biologic medicines. Take herceptin as an example. If a genetic test is required in order to identify potential responders to the drug, its cost will have to be included in the economic analysis.

Geographical considerations are equally important. A launch in the UK National Health System for example may require submission to NICE. NICE requires a threshold of cost-effectiveness below £30,000 before it will begin to reimburse products. The probability of achieving this threshold should be tested early in the product development process.

Critics argue early value-based pricing analyses based on pharmacoeconomics evidence are not based on "ideal" data as outlined above. The trick is not to rely on one model only, but to triangulate and link various models. Financial (ROI, NPV) models, health economics, and market forecast models should be tested. In other words, as clinical development progresses, pricing results from a preliminary cost-effectiveness threshold analysis should be tested in a market forecast model to see how it compares to previous assumptions that were perhaps based on financial targets only.

HEALTH ECONOMICS SUPPORT FOR COMPANY VALUATIONS AND LICENSING

Companies must merge the results from value-based pricing analysis based on preliminary cost-effectiveness models with the valuation model of a potential deal structure in licensing go/no go decision making. Financial valuation models for biotech companies include the risk of failing the clinical development stages while omitting substantial market risks after approval is granted like commercial risks (reimbursement). Commercial risks are related to the products potential cost-effectiveness as perceived by payors.

For example, an ambitious new product price that is not based on cost-effectiveness grounds (e.g. CE ≤ λ) may lead to overvaluations in the NPV (net present value) calculations. Another company buying the product may pay too much for in-licensing with market access problems arising later at launch. Overpaying for in-licensing, may lead to overvaluation of a biotech's pipeline which provide downstream disappointments in the stock market when expectations cannot be fulfilled. Therefore, a threshold minimum product price meeting the cost-effectiveness criterion should be calculated as the price for which the rNPV (risk adjusted net present value) calculation yields a positive value or is greater than the return of investment target value. This is an iterative process and should be updated along with the development progression and as soon as new data becomes available. As already stated in the pricing section before, the right combination of pharmacoeconomic data with traditional financial valuation methods such as rNPV will improve its accuracy and will be vital information in determining optimal deal terms and timing. Potential cost-effectiveness estimates become firmer as the compound progresses through the various development phases.

There is a growing range of techniques for accounting for marketing risk. The complexity of accurately predicting the development and market risks of dependence using one type of model suggest companies are using multiple models as a form of cross validation. The difficulty of predicting these risks also argues for an increased awareness on the part of payors of the structure and use of valuation models by the industry. The decisions payors are making on pricing and reimbursement today will reverberate into the future. Their decisions impact not just what drugs will remain in the market, but the input criteria used for drug valuations that will come to market in ten or more years.

Marketing the economic value proposition

Health economics has often been considered a marketing gimmick. Past methods and standards were less developed than today's standards. Some studies criticized the value of drug evaluations.

Value-based marketing strategies should entail much more than a back-of-the-envelope study. It should also contain more than the standard congress poster or slide kit. It should be a very comprehensive and evidence based communication instrument in the hands of company customer representatives.

Marketing an economic and health outcome based value proposition is about a consistent payor and market/formulary access strategy. The strategy should ensure patients needing the drug will have access to it. Having access to a new medical technology generally means having someone else pay for it. Biotech managers need to help their clients demonstrate the real economic value. Making products affordable may require creative, novel and sometimes even risk-sharing without the agreements. This is seen more and more often in oncologic drugs while NICE is still evaluating the drug for cost-effectiveness. Every biotech executive will need to form an opinion concerning the company's position on matters like pay-back clauses, conditional reimbursements, and risk-sharing agreements. The earlier the value demonstration component becomes part of a strategic marketing effort, the more effective it will be.

From a health economics perspective, the contributing factors of increased market access can be described:
- Product differentiation among increasingly economically sophisticated customers
- Economies of scale in marketing
- Market entry barriers set by dissemination of solid health economics evidence

Product differentiation

Increased competitiveness leads to increased efforts to differentiate products. These days, product differentiation is perhaps the most important contribution health economics can bring to the table to improve global market access.

Conventionally, drug companies marketed their products based on safety and efficacy. While these selling points undoubtedly remain central factors of differentiation, biotech companies are now being asked to distinguish their products on their overall economic value and their net impact on patients' well being/outcomes. The application of health economics early in clinical development can help identify which products are likely to meet this demand or provide insight into what data might be required for the demonstration of economic differentiation. An example, as one author put it[2]: Genentech "expanded the estimated market value of one of their products from $78 million to $350 million by incorporating an economic/quality of life component to a clinical trial and by consulting healthcare payers about what they wanted."

Economies of scale in marketing

An internal health economics capability will ensure strategies are developed which demonstrate the economic value of the product on a global basis (there is almost no one-fits-all solutions due to the different incentives in the various healthcare systems). The costs involved with these multinational economic studies are substantial. Fortunately, they are incurred only once. After a central health economics value dossier and model is developed, individual country adaptation can be performed requiring only a relatively moderate investment.

Furthermore, the cost incurred with the health economics add-on compared to a clinical trial's total costs are marginal. The additional study offers a unique opportunity to collect a standard data set in several countries at once. This allows easy and efficient development of localized evaluations as well as pricing and reimbursement/market access dossiers.

Market entry barriers

In economic theory several market entry barriers for new

---

[2] *Pharmacoeconomics and Outcome Assessment. A global issue. Euromed Communications Ltd. 1999:136.*

competitors and products are discussed. Barriers range from monopolistic market position or technology advantages to more 'soft' entry barriers like image, strong trademarks, or in the case of biotech drugs, product advantages supported by economic evidence.

Health economic evaluations demonstrating an excellent economic profile or improved quality of life may provide an entry barrier for newly developed, alternative products, and delay adoption or prevent success of competitor products.

Pharmacoeconomics can, in the same manner, create price and reimbursement hurdles to new market entrants by cost-conscious payors requiring demonstration of similar data and cost-effectiveness. Therefore, to be first to market with solid economic evidence may be key for market penetration and product uptake.

Using clinical trials for economic data creation

We already mentioned to piggy-backing economic analyses onto clinical trial programs wherever possible. A strategic marketing approach towards value demonstration should maximize clinical study opportunities. For example, Germany has set up a new agency, Institute for Quality and Efficiency in the Health Care System, (IQWIG) that will provide recommendations about future product reimbursements based on randomized clinical trial evidence. While the evaluation methods are in development, international standards of health technology assessment have widely been ignored. Economic modelling is currently not accepted at all by IQWIG. Companies considering Germany in their launch projections might want to think about including economic or patient reported outcomes into their clinical trials early on. In fact a company working on products whose main outcome will be a quality of live endpoint have mostly no other choice but to put a protocol in their phase III trial.

As previously discussed, in entirely new indications, good cost-effectiveness data will set a benchmark with which any follow-on competitor will be compared. This may serve as an important market entry barrier keeping away competitors for awhile. Thus it might be a wise move for medical, marketing and health economics executives at biotech companies to design studies with the payor/customer in mind early in a product's life cycle.

EDUCATION

It is important for biotech executives and key staff members to understand health economic concepts. The most sophisticated budget impact analyses and cost-effectiveness models are worthless if people cannot explain the reasons why they are doing it. New technologies and vendors have emerged allowing creation of user-friendly e-dossiers and value propositions as well as interactive sales force and e-learning tools. Focusing on key differentials of a drug is preferred to creating huge data dumps. Training sales reps and key account managers on the main concepts and value messages as well as role playing objection handling is likely to pay off in customer interactions.

ORGANIZATIONAL ASPECTS

The organizational execution of health economic inputs depends on the company's size and geographical activities. While large biotech companies such as Amgen and Genentech have established in-house capabilities, start-up and smaller companies might be doing well to hire one senior individual who can oversee all strategic health economics, pricing, and value demonstration/access matters. Over time, people than can be added to form a strong team.

IDEAL PROFILE OF THE HEAD OF HEALTH ECONOMICS

Given the biotech sector dynamics, a communicative self-starter, is needed. Ideally, the individual will have international experience (multilingual) who is also familiar with the relevant methods and the health policy environment. Biotech CEOs may want to be careful about not creating an in-house "university," but rather hiring an experienced yet hands-on and pragmatic manager who has the market knowledge to judge what really matters to customers, how to get things done best and quickly, and how to manage external vendors, academics and other consultants in order to get the best value.

Her/his key responsibilities should be:
- to create and disseminate product value propositions which align with product innovation targets.
- to integrate medical/scientific data and health economics research.
- to understand global stakeholder business environments and key stakeholders (wholesalers, payors/payor systems, governments, facilities, and physicians), geographies or customer segments.
- to develop and implement health economics initiatives to support products across their lifecycle.
- to provide input for product development ensuring inclusion of health economics output where possible.
- to provide health economics input to other functions such as investor relations/public affairs in order to influence policies and decisions.
- to join key negotiations where needed and to facilitate deal making and licensing as well as portfolio decision making.

This person should also be tasked with other activities like monitoring industry trends and anticipating future changes in the marketplace. The economics health specialist should take action to create long-term opportunities and value. Last but not least, he or she should be expected to build leadership and management depth by contributing to the strategic vision of the business through contribution to internal decision making and strategy discussions.

Success factors

Planning for the inclusion of economic evaluations into the development program of a new technology should begin at an early stage. Preliminary economic data should ideally be collected in phase II trials, and implemented into phase III trials and beyond. The development of an integrated value based development plan at these key-stages allows for the identification of some of the major reimbursement barriers to entry. The importance of a thorough economic package ready for negotiations with reimbursement agencies pre-launch and at launch cannot be overstressed.

To be successful in maximizing the advantages strategic pharmacoeconomics approaches can add to a biotech company, several factors have to be considered.

- First of all, early and continuous involvement with the head of health economics in the planning and prioritisation process is essential in order to ensure her/his input can contribute to decision-making.
- Health economics is a very interdisciplinary approach, depending on the input of various stakeholders within the company and their information supply. A strong hand-in-hand collaboration with clinical and business executives is a prerequisite for successful management of these projects. In early development phases, health economics may co-operate closely with business development to provide timely and relevant intelligence to facilitate evidence-based decision making on portfolio issues and in- or out-licensing opportunities.
- At a later stage, a joint effort between the health economics and clinical head is required to ensure a smooth data collection alongside the trial and to avoid delays. It also ensures the health economist has an impact on trial design and study location in order to be able to provide and collect relevant economic evidence.
- To be able to truly influence strategies and to obtain relevant and timely information on business strategy, project status, and development, the health economics head should be part of the senior

management team, and also be regular member of relevant project development team meetings. This provides an opportunity to educate the whole company concerning the application and the benefits of pharmacoeconomics, and to gain support from various key executives in the company.
- Small companies will need to outsource probably more than 80% of the technical work. Therefore a nonbureacratic and flexible budget for health economics should be available.
- Due to increasing external as well as internal requirements of health economics research, it is important to keep abreast with developments and align resources and capabilities to be able to meet the needs of all stakeholders.

CONCLUSIONS

Health economics is here to stay and requirements will increase globally. Biotech executives will have to consider additional dimensions of market access risk caused by increasing payor-side cost pressures. These must be taken seriously to remain competitive within the industry. Executives must also shape their organizations in a way to meet these challenges. The consequent adaptation of the concepts outlined in this chapter will play an essential part towards a payor-focused business strategy.

Valuation of Technologies Through Services
Ingrid Marchal-Gerez

Ingrid Marchal-Gerez is business development manager at Protéus, a French biotechnology company focused on the use of proteins as therapeutics and manufacturing tools. Ingrid can be contacted at *ingrid.marchal@gmail.com*.

Many biotech companies have their origin in the development of a significant technology in a public lab. Genentech, arguably the first biotech company in history, stems from the application of the first genetic engineering techniques back in the 1970s.

For a biotech venture, there are several ways to exploit key technologies: one is to apply the technologies in internal, proprietary research, with the goal of developing new drugs, as did Genentech. This strategy is risky and costly, but, if successful, highly rewarding. Companies founded with this model need to go through several large financing rounds to fuel their R&D because they are not likely to get income before years of development.

An alternative solution is to commercialize the technology and work with other companies.

This can be done through research collaboration, where both partners share the risks and the success of their research. Generally, the biotech company receives payments from its industrial partners at the launch of the project and when certain milestones are reached. The financing partner retains the property on potential compounds and, when products are marketed, royalties are paid to the biotech company.

Another possibility is to offer service, based on your technology, and derive service charges. Although it might seem less rewarding, this model also has its advantages: it is easy to initiate and can rapidly generate income and, hence, reassure investors. It also helps starting companies to structure themselves and acquire professional, industrial practices. Finally, agreements based on service can build a company's credibility and reputation, and ultimately lead to strategic collaborations once there is enough confidence between the partners.

Service can be defined as supplying of a precise piece of work in exchange for a certain amount of money, the service fees. Service agreements generally exclude the payment of any royalties, which are frequently involved in more global collaborations or alliances. A variety of activities can be outsourced through services, including molecular biology (sequencing, cloning, gene synthesis); chemistry (modelization and design, lead optimization, library synthesis); biochemistry (protein expression and analysis); drug development (preclinical and clinical studies); manufacturing (protein or compound production); supply chain management services, and others.

This section will describe how and why commercializing service can be a viable and interesting start for a biotech company in balancing risks and building solid bases for the future. It will also approach how to succeed in such a strategy and how to avoid the traps inherent in this model.

BIOTECHNOLOGY SERVICES: AN UNDERESTIMATED, THOUGH INTERESTING BUSINESS MODEL

By offering service, a company grants a limited access to its resources (i.e. proprietary technologies, know-how, platforms, workforce), in a limited time frame, for a defined objective. The drawback is that subsidies derived from services are limited, so the company has to rely on a large client base. However, because there is a demand for outsourcing and service supply, this model can make economic sense.

WANTED: INNOVATIVE TECHNOLOGIES

Biotech companies can be highly inventive and are responsible for smart enabling technologies. Inventors generally believe in their own technology; the good news is that others might also realize its potential.

Pharma is recurrently said to suffer from a lack of innovation. Companies are urged to find new molecules, new applications, new tools to keep them ahead of their competition. Any new technology that might help them accelerate drug discovery, validation, development, delivery or even production is of potential interest. All big pharma firms have dedicated teams whose role is to identify new molecules and technologies, and your technology might be their next blue chip.

The same is also true for other fields of application, e.g. in diagnostics, bioenergy, functional foods. All are major players actively searching for new technologies that will make a difference in their market.

Protéus, a French biotech established in 1998, specializes in the discovery and development of novel proteins and protein-derived processes and products for industry. From inception, Protéus' strategy was to specialize in a set of enabling technologies rather than in a therapeutic or industrial domain, and, rather than products, has developed excellence and know-how that are now seen as helpful resources for other industrial or healthcare companies.

The company has built an integrated platform with four major components:

(i) a broad and original collection of extremophile microorganisms
(ii) high-throughput functional screening systems including robotics, clean rooms, and patented enzyme substrates
(iii) protein optimization based on two complementary patented technologies, EvoSight™ for rational directed evolution and L-Shuffling™ for gene shuffling
(iv) proprietary tools for small to large scale expression of proteins and process development

The strategy of Protéus has been to establish long-term partnerships with major industrial players such as Degussa, Henkel,

Aventis or DSM, with the objective of developing new enzymes for industrial processes. Successes obtained in these programs have positioned Protéus as a key specialist in the field of protein discovery and engineering.

Under these partnerships, Protéus uses its technology platform to mine its biodiversity collection for a sought enzyme activity, adapts the enzyme to the process constraints, and develops a scalable industrial production system. The agreements combine access fees, research fees, milestone payments and finally royalties.

Some companies expressed a need for accessing the technologies in a more flexible manner, and for application to their own proteins. Moreover, some customers were interested in using only part of Protéus' capabilities rather than the complete platform. To address this repeated demand, Protéus started providing services and formed a specific division called PSI (Protéus Services for Industry). According to Protéus' CEO and founder Daniel Dupret, PSI does not compete with the research partnership collaboration mode; rather, it offers an alternative way of offering Protéus' technologies, allowing the company to meet a larger range of needs, and collaborate with smaller as well as with larger corporations.

As stated by Dr. Gary Pisano at the Harvard School of Business, "one tool doesn't do it, but the integration of tools can certainly affect the way drugs are discovered, refined, developed or produced." There is clearly a demand from large companies for accessing exclusive technologies together with high level expertise and know-how, on a project-by-project basis. Companies that can respond to this demand, and apply their technologies to other firms' projects are likely to succeed.

The Rise of "Virtual" Companies

Besides traditional, integrated pharmaceutical companies, there is an increasing trend for the creation of small companies with high capitalization but little personnel. These "virtual companies" need to outsource a major part of their operations, and thus demand biotech services.

Because developing drugs is an increasingly expensive and

risky business, some biotech firms adopt a virtual company model to reduce their risks. These companies have chosen to focus on the development of their molecules, and hence have secured important financing rounds. Yet, rather than consuming time and money in building their own permanent labs and research teams, they allocate their resources on demand through outsourcing. The benefit is double: first, they have access in a timely manner to fully efficient equipment and personnel; second, they limit their own risks by being flexible. In case of necessity, it is much easier to stop a service than to cut down a whole research department.

For these companies, service is an attractive collaboration model: negotiations are usually simple, and the agreements are less committing than research partnerships.

An example of such strategy is the Franco-American company Cerenis. The company was founded in 2005 by experienced pharma executives with the aim of developing novel HDL (high density lipoprotein) based therapies in cardiovascular and metabolic diseases. Cerenis is financed by venture capital: the company successively raised $30 million in 2005 and $53.3 million a year later.

Cerenis is based both in the USA at Ann Arbor, MI, and in Toulouse, France. Rather than investing in its own labs, Cerenis uses some external capabilities and makes a broad use of outsourcing, thus focusing its own efforts on the development of drugs. The importance of funds raised so far reflects the confidence of investors in this strategy.

This virtual company strategy has its own risks, but it can be a sensible way to perform initial proof-of-principle demonstration studies and delay the search for external funding while increasing the valuation of a company. It also stimulates the market of biotech service companies, from basic (e.g. sequencing) services to much more sophisticated ones. Biotech companies that are able to offer exclusive and effective technologies through simple and easy to handle service agreements are thus likely to find opportunities with these drug development-only firms.

Revenue From Service is Better Than No Revenue

An obvious benefit of offering service based on your technology is that it generates revenues. Developing a cutting-edge technology costs time and money, but developing new products costs much more, and the rewards come years later. The fruits can, however, be reaped more rapidly by making proprietary technologies available to others, either through licensing, research partnerships, or services.

Research partnerships generally involve lengthy negotiations in which the amount of milestones and royalties is thoroughly discussed. In the long term, royalties received on the derived products, if commercialized, may be lucrative, but this is associated with a high risk level.

Offering service in exchange for service charges is simpler and generally more suitable if the service is a well-defined one, when the time and effort can be estimated in advance. The profit derived from selling a service is smaller, but will come earlier, with virtually no risk.

Many early-stage companies engaged in drug discovery and development adopt a hybrid strategy, combining service offering and proprietary research, in order to balance their risks.

Revenues generated through services can finance at least part of the R&D burden and help increase the company's stability and longevity. It can also reassure employees in the value of their technologies and work, and act as an incentive to stay with the company, a factor that must not be neglected.

Evotec is a leader in the discovery and development of novel small molecule drugs. The company is the result of the merger in 2000 of Evotec AG (founded in 1993 in Germany) and OAI (founded in 1992 in the UK). Evotec is publicly traded on the Frankfurt Stock Exchange.

Evotec entered the market as a technology-focused company, collaborating with pharmaceutical firms to develop new tools and technologies to improve the drug discovery process. These capabilities are integrated into one of the most advanced discovery engines available on the market.

Today Evotec has a hybrid strategy, combining service and proprietary or partnered research. The services division offers integrated solutions from drug target to clinic, while the pharma division focuses on finding new treatments for diseases of the central nervous system (CNS). According to Anne Hennecke, Senior Vice President Investor Relations & Corporate Communications, offering services allows Evotec to sustain and continuously improve their discovery engine. It also covers part of Evotec's proprietary research expenses; only the clinical development requires external funding.

While most biotech companies have the ambition to bring new life-saving therapies to the market, it is a known fact that few will succeed. Other ways of generating their income, such as offering their technologies and know-how through services, are therefore worth consideration, and are necessary to build a global balanced biotech market.

STRUCTURING A BIOTECH COMPANY THROUGH SERVICES

Besides favoring the economic stability of biotech companies, service can shape them into mature, industrial partners.

TECHNOLOGY VALIDATION

Most researchers are confident in the power of their technologies and tend to overstate their supposed influence in the discovery process. Yet it is only by confronting their techniques and know-how in competing technologies, and by exchanging with external potential users that they can really evaluate the utility of their inventions.

A technique is generally powerful in the precise context where it has been developed, and for the precise issues it was meant to address. The versatility and robustness of such technology is not granted until it is successfully applied in a variety of different contexts.

Offering service thus has this additional advantage in demonstrating the power of proprietary techniques and hence the solidity of a company's scientific foundations.

In the case of Evotec, providing services rapidly confirmed the capabilities of the company and the broad applicability of the technologies. In return, this encouraged more customers to work with Evotec, and stimulated investors to support the development of the company.

Any new entrant in the biotech field has to show evidence and prove the value of their tools and products. Offering service is an effective way to validate the technologies and demonstrate their flexibility, reproducibility and usefulness.

BUILDING A STRONG IP PORTFOLIO

Besides strengthening the confidence in techniques themselves, a company that intends to market proprietary technologies needs to consolidate its intellectual property to avoid any counterfeiting issue and reassure customers.

As in any research-intensive industry, where the cost of innovation is high and the cost of imitation relatively low, intellectual property protection is essential for biotechnology. Only strong IP protection can secure a competitive advantage. Without appropriate protection, a company could see a competitor exploit the same technologies and offer lower prices, reflecting its reduced R&D costs.

Patents can protect key technologies and products. A patent grants the applicant the right to prohibit competitors from practicing and commercializing the invention it describes. However, a patent does not grant the right to practice the invention, as it can be dependent upon third parties' intellectual property. To secure their invention, inventors must not only file a patent, but also meticulously demonstrate their freedom to operate, i.e. their independence of other's patents. They should also scrutinize the prior art to ensure the invention they describe really is new. Failing such evidence, it would be too easy for competition to challenge the patent.

An efficient intellectual property policy has to focus on all these aspects, i.e. bringing evidence that the technology is independent and new; filing and securing valid patents; as well as ensuring it is not counterfeited by any other player.

At Protéus, IP has always been considered as integral to the company as the technologies themselves. As Protéus operates in the highly competitive field of biodiversity mining and directed evolution, it was essential to build a solid IP portfolio. Protéus has filed over 100 patent applications. Some have been granted on key technologies, including L-Shuffling™, a gene shuffling technique allowing the optimization of proteins such as enzymes and therapeutic proteins, and Phenomics™, an *in vitro* protein expression technique, confirming their originality. Surveying competitive technologies is also essential to sustain the company's freedom to operate.

When the company started its commercial operations and established research alliances with industrial players, the partners performed due diligence studies on Protéus's technologies, providing further evidence that Proteus's IP is strong and independent of external technologies.

Today Protéus has diversified its collaboration opportunities by offering contract research services based on its proprietary technologies. Being able to address its customers' concerns in terms of IP is a strong asset for the company, and is at least as important as the effectiveness of the technologies and the project management style.

When a company outsources a piece of work, even through service, it has to ensure no third party's patent is infringed by the service. Biotechnology service providers have to reassure their customers, not only on the scientific side, but also on potential legal implications of their collaboration. Offering service thus forces biotech companies to question the validity and solidity of their IP and to set up a strong IP protection strategy, an asset that will be beneficial at later stages of development.

Acquiring Industrial Methods

More generally, a young biotech company seeking to collaborate with larger firms through services or research partnership has to adapt to an industrial environment, which involves acquiring new habits and practices. Service providers have to adopt consistent methods, procedures and quality control, and train their employees in these industrial practices.

Such practices include compliance with predefined procedures and protocols, systematic recording of all operations, and the existence of clear project management and quality control policies. Speaking the same language and being able to provide documented evidence of what is done under the service agreement is indeed essential to gain and maintain customers' confidence.

Compliance with recognized policies such as international Good Laboratory Practice guidelines or International Organization for Standardization (ISO) standards is highly appreciated. Obtaining a certification of compliance is not necessary but is of course seen as strong evidence of quality.

Hybrigenics is a French biotechnology company founded in 1997 and specializes in protein-protein interaction mapping. This expertise has been valued in several ways. In its early days, the company commercialized protein interaction maps generated by proprietary research, while pursuing partnered research projects aimed at discovering new therapeutic targets. Today Hybrigenics' mission is to advance the development of its small molecule drug candidates, particularly in oncology. Besides this research and development activity, Hybrigenics provides access to its functional yeast two-hybrid platform on a fee-for-service basis. The company has built a solid reputation in the field of interaction mapping and pathway analysis. In addition to its solid scientific background, Hybrigenics puts a strong emphasis on the quality and consistency of its service. The company received an ISO 9001:2000 certification for its yeast two-hybrid service, now recognized as the most advanced and reliable interaction mapping service available to date.

By working with industrial firms, young biotech companies can benefit from their experience, learn and be trained in industrial methods and practices. This early interaction contributes to shaping sustainable businesses by favoring awareness in procedures and regulations.

Market Yourself

Another advantage of building early relationships with industrial firms is that biotech companies have to learn how to market themselves.

Some young biotech companies pay little attention to marketing and communication because they are focused on research. Yet it is the combination of strong science and effective communication that will attract the interest of potential partners and investors. Defining an effective marketing strategy early on is thus an advantage even for young companies.

Companies that offer service, however, need to hit a large client base and hence must make intensive marketing efforts.

In large firms, people are overwhelmed with commercial information from biotech service providers worldwide. A company that wants to promote its own product or service has to distinguish itself with an aggressive and well-targeted marketing strategy; the technology will not do it all. The aim is to influence the buyers, so that they think about your service first.

With regard to its service marketing strategy, Hybrigenics has been exemplary. From advertisement in commonly read papers to the utilization of sponsored web links, Hybrigenics has been able to attract the largest part of scientists looking for yeast two-hybrid solutions and to gain customers all over the world. The company realized that no matter how good the science is, it must be publicized before it gets widely known and applied. This efficient integration of marketing practices has allowed Hybrigenics to rapidly increase turnover derived from services from 150 000 € in 2003 to 2,4 M€ in 2006—1600% growth over three years.

Selling services implies that the selling company must be ready to go to the market, even if it is newly created. Companies employing this business model tend to mature and acquire professional practices early on for all aspects of their business, including IP, procedures and quality controls, and marketing. These companies are therefore well prepared for their later stages of development.

OPENING DOORS

Biotech service companies have an early interaction with a number of customers, including big firms. Besides influencing their maturation, this can also lead to new opportunities and put a company on the track to realize its deeper ambitions such as product development.

EXTEND YOUR MARKET

Have you ever thought of all the applications your technology could have? It probably goes far beyond your own field of expertise, and interacting with external partners or customers might open your eyes to a whole new host of applications, even in remote areas.

When Protéus developed its versatile protein expression platform, the objective was to express enzyme libraries in order to test their function and rapidly screen for the variants with desirable properties. In particular, Protéus developed a proprietary *in vitro* protein expression technology, Phenomics™. This technology has been largely applied in proprietary and partnered research programs and proven to be robust and efficient for functional expression of a variety of proteins.

When Protéus launched its service activity, management realized the technology was of high interest in drug discovery activity, because it also allows the rapid expression of target proteins that are functional and suitable for high throughput screening of compound libraries.

While the company's original field of expertise is more relevant to industrial (white) biotechnology, offering service led it to penetrate the drug discovery market and obtain important and repeated successes in this field.

Another example is Evotec, which specializes in central nervous system therapies. Offering service allows it to be active in a number of other therapeutic areas, where Evotec has no particular expertise.

Selling a proprietary technology through service can therefore be seen as a way to broaden its scientific horizon, acquire further expertise and, in some cases, evolve from its original field to a more open, promising one.

BUILD A SUCCESS STORY

Some companies that started to sell their technology as a service are now well-established biopharmaceutical firms with a pipeline of their own and a variety of partnered products.

Crucell is a publicly-traded company headquartered in Leiden, Netherlands. When the company started in 1993 as Introgene, it was focused on building technologies and tools for gene-based therapy, in particular viral vectors. In this context, Introgene had a first significant partnership with Genzyme. In 1995, Introgene developed the PER.C6® cell-line, a human cell-line first used for the production of viral vectors; the technology was commercialized through production services. The range of applications rapidly expanded to the preparation of vaccines, therapeutic proteins and monoclonal antibodies. The success of the technology has since been confirmed by licensing agreements with various biopharma companies, including Merck, Sanofi-Aventis and Centocor.

The company has also teamed up with DSM to create joint PER.C6 research and production centers in Europe and in the USA. Having built a profitable business thanks to its technology, Crucell has been focusing on the development of its own pipeline of vaccines against infectious diseases.

Starting with contract research services, Crucell turned into a solid, diversified business. The technologies are now recognized and made available to partners through either licensing, services or research alliances. The company also pursues its proprietary vaccine and drug discovery activities, building a pipeline to ensure its long-term growth.

This illustrates how, by combining valuable technologies and a wise business development, a company can rely on technology services as a growth lever. It also shows that, in some occasions, a satisfied customer may decide to extend the relationship and enter into strategic collaboration with the company.

Biotech companies can thus use service offering as a first step to fuel their growth and develop other, more lucrative but more expensive activities such as internal research. For an established company, it can also be the first step into a new market, as Protéus is doing for drug discovery.

CONCLUSION

Searching for new life-saving or life-enhancing products is the ultimate goal of the biotechnology industry. This research and development activity is extremely expensive, and no profit is likely to come for years, if ever. The consequence is that the whole industry is extremely investment-consuming: according to BIO, the Biotechnology Industry Organization, $20 billion was injected in biotechnology in the USA for 2004. Venture capital accounts for about a quarter of all investments.

However, not all biotech companies are directly involved in pharmaceutical research. The market is effectively balanced between research product suppliers, service providers and research-oriented companies, as well as companies using a hybrid model.

There is a strong need for high tech, high quality service providers. For companies focused on technology rather than on products, it is an effective way of valuing their proprietary assets and earning money. Providing services helps the companies to mature with minimal risks and be prepared for the future stages of development. As exemplified by some success stories, providing service can be the first piece of a global strategy that will allow the company to stand out in its market.

Of course service is not the only or the best way to prosper. When a company raises funds from investors, it should focus its resources into research towards discovering new products. Yet, service should be seen as a possible business tool that can provide new opportunities by extending the market, and can provide additional monies to fuel the research.

How Big is the World Market for Biopharmaceuticals? It Depends on How You Define Biopharmaceutical
Ronald A. Rader

Ronald A. Rader is president of the Biotechnology Information Institute and the author of Biopharmaceutical Products in the U.S. and European Markets, the only reference book/database concerning biopharmaceuticals. Ran can be contacted at *biotech@biopharma.com*.

Any review of recent presentations and publications, including the trade, scientific and business press and industry studies, will show a wide range of estimates for size of the biopharmaceutical product world market. Estimates of the current market commonly range from $40-$60 billion to about $80 billion or more. Most of these discrepancies are due to differences in what is and what is not included as being a biopharmaceutical. The biopharmaceutical industry is a relatively mature industry with about 300 products in the U.S. market, including over 100 recombinant protein products. Despite this and the strategic importance of the industry, anarchy prevails regarding biopharmaceutical classification and the industry's basic identity. This division affects estimates of total industry size/revenue and detracts from an understanding and analysis of the industry.

FOUR VIEWS/PARADIGMS OF BIOPHARMACEUTICAL

Four basic, often conflicting, definitions of biopharmaceutical in common use[1,2] are:

> **Broad Biotech**: This entity or active agent-based definition, which is the predominant definition within the U.S. biopharmaceutical industry, views biopharmaceuticals as pharmaceuticals inherently

biological in nature due to their manufacture using live organisms (biotechnology, bioprocessing). This includes not just recombinant proteins and monoclonal antibodies, but also vaccines, blood/plasma products, non-recombinant proteins, gene therapies, and cultured cellular and tissue products. In contrast, most pharmaceuticals are drugs: i.e., pharmaceuticals are inherently chemical (not biological) in nature manufactured by chemical (not biotechnological) methods. From this view, small molecules and other drugs are not biopharmaceuticals. Worldwide biopharmaceutical revenue in 2006 was about $89 billion, or less than 15% of the total pharmaceutical market.

New Biotechnology: This second entity-based definition takes a subset of the broad view, restricting biopharmaceuticals to products incorporating "new" genetically engineering biotechnologies, primarily including just recombinant proteins and monoclonal antibodies. This view is more prevalent in Europe and is in line with European Union (EU) regulations concerning "biotechnology medicinal products," and reflects the current dominance of the industry by recombinant protein and monoclonal antibody products (discussed below). This definition is somewhat arbitrary or flawed, in the sense that among the excluded products are many products that have been developed and approved in recent decades incorporating more recent and more complex technology compared to many recombinant proteins and monoclonal antibodies. These technologies may now deserve to be considered 'old' because they were invented in the 1970s and commercialized in the 1980s. From this view, worldwide biopharmaceutical revenue in 2006 was about $65 billion.

Biotech Business: This company business model- or image-based definition considers biopharmaceuticals as all pharmaceuticals involving biotech-like (smaller, entrepreneurial, R&D-intensive) companies or high-tech companies, and also includes obvious biopharmaceuticals from large companies (Big Pharma). Products, technologies and companies need not have any involvement with or use of biotechnology to be labeled biopharmaceutical. Biopharmaceutical (and biotechnology) are often used as metaphors to describe any/all life sciences or pharmaceutical R&D, e.g., drug discovery and screening. Thus, most or even all small molecule drugs, technologies and companies are considered or advertised as being biopharmaceutical (and biotechnological). Those commonly using this definition include many in the popular and financial press, stock analysts, and the Biotechnology Industry Organization (BIO). This definition is popular in many of the most respected, presumed authoritative industry annual and other analyses, including those from financial and accounting firms, stock brokerages, and market research studies. Users of this view generally claim to know a biotech/biopharmaceutical product, technology and company when they see it (if is seems high tech and involves pharmaceuticals, it's biopharmaceutical), and perceive lack of clear criteria as reflecting ever-evolving perceptions of what is/isn't biotechnology. With this view rarely, if ever, defined by those using it (this author has never seen it fully defined by its users), and with subjective, defining criteria varying from year-to-year and analyst-to-analyst, much of what has been and is written about market size and trends in the biopharmaceutical industry should be viewed as highly suspect.

Pharma Business: This second company image-centric or business model-based definition simply includes all pharmaceuticals as biopharmaceuticals, i.e., biopharmaceutical is used as a synonym for pharmaceutical; and the pharmaceutical industry is now the biopharmaceutical industry. This term's use is traceable or linked to what this author views as the "myth of convergence," and associated with pharmaceutical/drug industry efforts to rebrand itself as biopharmaceutical (which sounds better than pharmaceutical or drug). Various industry analytical reports, including those funded by the Pharmaceutical Research and Manufacturers Association (PhRMA), assert the pharmaceutical industry (particularly, Big Pharma) has recently morphed or transformed itself into the biopharmaceutical industry as a result of the "convergence" of biotechnology and pharmaceutical industries/technologies. That is, the pharmaceutical industry, through extensive adoption of new (bio)technologies and close relationships with biotech companies (primarily outsourcing and licensing), has undergone a fundamental change. With this transformation or convergence, everything in the pharmaceutical universe is now biopharmaceutical. From this view, (bio)pharmaceutical revenues in 2006 were about $650 billion.

The two business/image-based biotech and pharma business paradigms, in this author's view, are primarily based on perceptions and portrayals (and often hype) about products and companies classified as high tech, and simply do not lend themselves to consistent analysis of biopharmaceutical sales/revenues or other aspects of the industry. These views are subjective and undefined. Inclusion of small molecules and other drugs (chemical-based, non-biological pharmaceuticals), and drug discovery and other service companies with no involvement or use of actual biotechnology negates the utility of many of these analyses (in terms of characterizing actual biopharmaceutical products and the industry). Arbitrary inclusion of small molecule drugs considered biopharmaceuticals can be readily seen in the lists/reviews of biopharmaceutical approvals issued by BIO, PhRMA and others. Inclusion is also seen in business and popular press articles, and in many of the presumed authoritative annual reviews of the biotechnology and pharmaceutical industries issued by financial, accounting and other firms. These views may be useful to bring together life sciences and pharmaceutical companies, but the biopharmaceutical (and biotechnology) label is inaccurate. For many users of the Biotech Business view, the ability to classify products, companies, technologies and the industry as being biotech/biopharmaceutical is based on perceptions. What's in vogue, how it's hyped, etc., is all-important. From their view, biotechnology is an ever-changing business model, or even a state-of-mind no longer defined by actual (bio)technologies. For many companies and investors, their main interests lie with stock and company valuations, compelling R&D stories, public image, etc. Loose and ever-evolving definitions lend themselves well to these interests.

The pharmaceutical industry has not transformed or morphed into the biopharmaceutical industry contrary to the Pharma Business view. Biotechnology and biopharmaceuticals, whether considered by products, R&D expenditures, company revenue or any other relevant parameter, remain a small portion (~15%) of the pharmaceutical industry. The number, percentage and contributions of biopharmaceuticals relative to drugs and all pharmaceuticals is growing, but biopharmaceuticals remain a distinct and small subset of the pharmaceutical industry.

Despite the growth, newer R&D methods, R&D outsourcing, and Big Pharma in-licensing, the underlying nature of the industry, which remains predominantly concerned with small molecule drugs (not biopharmaceuticals), has not changed. Taken as a whole, pharmaceutical companies, including Big Pharma, have only a small involvement percentage with biopharmaceutical products. And, biotechnology methods have always been fully integrated into pharmaceutical/drug R&D, e.g., all *in vitro* and *in vivo* testing is considered biotechnology. The drug screening, structure-activity relationships, and other data-driven technologies hyped as biotechnology have long been in use (adopted in the 1970s). Industries are defined by their products and related methods of manufacture, not by their research methods. Companies making steel or automobiles are not labeled as part of the metallurgical industry. Renaming the pharmaceutical industry to be biopharmaceutical is comparable to proclaiming (as some authors have) the biotechnology and pharmaceutical industries are now part of the nanotechnology industry, since their R&D, technologies and products are all at the molecular or microscopic scale.

Besides including non-biopharmaceuticals, business/image-based industry analyses often oversimplify and classify a whole company (and its revenue) as 'biopharmaceutical' instead of summing revenue data for individual products. At one time—in the 1980s—this yielded valid analyses. However, today it is difficult or impossible to determine company revenues from biopharmaceuticals vs. drugs vs. other sources. Excepting the smallest companies, most companies involved with biopharmaceuticals are now primarily involved with drugs and/or have other revenue sources.

Estimates for worldwide 2006 biopharmaceutical product revenues by major product classes are shown in Table 1. These data are based on the Broad Biotech view of what is a biopharmaceutical (see above), as shown by the various product classes included. These estimates are based on product revenue as reported in *Biopharmaceutical Products in the U.S. and European Markets*[3], and primarily involve summation of 2006 product revenues (total sales) as reported by the primary manufacturer(s)/marketer(s) or the author's estimates where these data were unavailable. For those products/industries, notably vaccines and blood/plasma products, dominated by just a few large companies not disclosing product-specific revenue, consensus estimates of industry-wide revenues are used. To calculate totals, products are included in only one class, which mostly affects vaccines. Recombinant Hepatitis B vaccines, the only recombinant vaccines with significant revenue, account for an estimated $1.5 billion and are included with Recombinant Proteins. These data do not include biotechnology-based diagnostics (except for a few radioimmune diagnostics administered systemically). Reliable data were generally available, particularly for products with the most sales. Note, these data do not reflect total industry revenue, i.e., they do not include total end consumer or pharmacy sales, sales by middlemen, sales of intermediates, contract manufacturing or patent royalties.

The $93 billion total includes $1 billion for the unknown num-

Table 1: 2006 worldwide biopharmaceutical revenue by product class

Product Class	Sales ($ millions)
Recombinant proteins (rDNA)	65,300*
Monoclonal antibodies, rDNA	19,500
Insulin Products (nearly all rDNA)	8,300
Monoclonal antibodies, non-rDNA	300*
Vaccines	9,500
Vaccines, non-rDNA	8,000*
Enzymes, non-rDNA	500*
Toxins (Botulinum)	1,050*
Cultured cells/tissues	100*
Blood Products (human and animal)	15,000*
Plasma-derived proteins	9,300
Cellular components	5,700
Misc. foreign biogenerics, knock-offs, etc.	1,000*
Total	**~$93,300**
* Indicates number used in total	

ber—easily hundreds—of biogenerics in the world market (probably more than all biopharmaceuticals in the U.S. and European markets).

Most do not realize copies of just about every successful biopharmaceutical originally developed by U.S. and European innovator companies is already manufactured and marketed in many other countries, essentially wherever lack of enforcement of granted patents allows them to be sold. In terms of numbers, most biopharmaceuticals in the world market are biogeneric, follow-on protein, biosimilar, or knock-off products.[4,5] For example, in China (PRC) alone, there are more than 20 manufacturers of recombinant granulocyte-colony stimulating factor (G-CSF) copies of Leukine from Amgen, and there are more than 180 insulin products reported to exist in the world market. With no reliable sources of information about these products/companies and with some of these biogenerics, e.g., various vaccines, manufactured by governments or their proxies, reliable revenue estimates do not exist. Many simply ignore these foreign product contributions. This is in part due to Western bias, the reality that the markets for these individual products is small, the fact these products do not affect markets for innovator products, and also because biogeneric approvals are often based on lower standards.

From the Broad Biotech view, world biopharmaceutical revenue is about $93 billion. If one takes the New Biotech view or otherwise just considers genetic engineering, primarily recombinant protein and monoclonal antibody products, total industry 2006 revenue is about $65 billion. This figure accounts for 70% of total biopharmaceutical revenue (taking the Broad Biotech view). This New Biotech view and frequent use of older data is the source and reason for the $40-50 billion/year reports of biopharmaceutical industry revenue. The estimated $20 billion market for monoclonal antibodies is now almost entirely comprised of recombinant products, with few non-recombinant products even on the market. Depending on one's biases, other product classes may be excluded or ignored. For example, some exclude various biologics, e.g., leukocytes and red blood cells, produced by local blood banks worldwide as biopharmaceuticals; and some even ignore vaccines.

The total worldwide pharmaceutical market in 2006 was estimated to be $643 billion, up about 7% from 2005.[6] At best, biopharmaceuticals comprise about 15% of the world market for pharmaceuticals (i.e., $93.3 billion/$643 billion). This is in line with other analyses reporting biopharmaceuticals comprise about 15% of the world pharmaceutical market. For additional perspective, total 2005 pharmaceutical revenues for Pfizer were $44.3 billion; GlaxoSmithKline, $34.0 billion; and Sanofi-Aventis, $32.3 billion. Among the largest biopharmaceutical companies, Amgen had $12.0 billion in revenue, and Genentech, $5.5 billion.

Biopharmaceutical product sales and product launches are increasing 15%-20% annually. This is twice the rate of the overall pharmaceutical industry. Therefore, it may be valid to claim, in the context of biopharmaceuticals vs. other pharmaceuticals (drugs), that biopharmaceuticals are the fastest-growing sector of the pharmaceutical industry. It is also commonly reported that pharmaceutical industry revenue, particularly large international pharmaceutical companies' revenue is, and will increasingly come from, new, innovative biopharmaceutical products. Much of this is due to a large number of recombinant monoclonal antibodies currently in development, with a number of these expected to be blockbusters or achieve otherwise healthy sales.

Prices for biopharmaceuticals vary over an incredible range. Some, e.g., enzymes for treatment of genetic enzyme deficiencies and some monoclonal antibodies for cancer treatment, cost over $100,000 annually for a course of treatment. Others, such as some pediatric vaccines purchased in bulk by governments and public health organizations cost only a few dollars or even cents per dose.

HOW LARGE IS THE U.S. BIOPHARMACEUTICAL MARKET?

The U.S. constitutes the largest market for biopharmaceuticals and all pharmaceuticals, and generally accounts for over half of all biopharmaceutical sales. As reported by IMS for 2006, North America accounted for $290.1 billion or 45% of global pharmaceutical revenue. Presuming a comparable 45% of biopharmaceutical sales are in the U.S., current U.S. annual biopharmaceutical sales are about $40 billion. However, biopharmaceutical sales are disproportionately higher in the U.S. (and other affluent major market countries), and the U.S. likely accounts for 50% or more of current worldwide biopharmaceutical sales or about $45 billion/year. Biopharmaceuticals tend to be much more expensive than most other pharmaceuticals; and their sales tend to be concentrated in more affluent, highly developed countries, particularly the classic, major pharmaceutical markets of U.S., Europe and Japan. Biopharmaceuticals are developed, launched, and established first in the U.S. market. With few exceptions, just about all biopharmaceuticals with significant sales are marketed in the U.S. This is particularly true for recombinant proteins and newer products. Thus, when considering biopharmaceuticals relative to drugs and all pharmaceuticals, the U.S. and other major market countries account for a disproportionately higher share of overall pharmaceutical sales.

The markets for biopharmaceuticals in lesser-developed countries, e.g., India and China, are rapidly growing. However, much of this involves locally-produced biogeneric copies of products originally developed by U.S. and European innovators (wherever lack of patents or their enforcement allows), with many of these products sold at low prices (relative to the original/innovator products in the U.S. and other major markets), and with the great majority of these products having just a few or, at best, tens of millions of dollars in total annual sales. However, with growing affluence, sales of biopharmaceuticals are increasing faster in many lesser-developed than in major, affluent markets.

REFERENCES:
1. Rader, R.A., What is a Biopharmaceutical? Part 1: (Bio)Technology-Based Definitions, BioExecutive Intl., March 2005, p. 60-65.
2. Rader, R.A., What is a Biopharmaceutical? Part 2: Company and

Industry Definitions, BioExecutive Intl., May 2005, p. 42-49.
3. Rader, R.A., Biopharmaceutical Products in the U.S. and European Markets, BioPlan Associates. Sept. 2007.
4. Rader, R.A., "What Is a Generic Biopharmaceutical? Biogeneric? Follow-On Protein? Biosimilar? Follow-On Biologic? Part 1: Introduction and Basic Paradigms," BioProcess International, p. 28-38, March 2007.
5. Rader, R.A., "What Is a Generic Biopharmaceutical? Biogeneric? Follow-On Protein? Biosimilar? Follow-On Biologic? Part II: Information, Nomenclature, Perceptions and the Market," p. 20-28, May 2007.
6. "IMS Health Reports Global Pharmaceutical Market Grew 7.0 Percent in 2006, to $643 Billion," IMS Health press release, March 20, 2007

About the Editor

Yali Friedman holds a doctorate in biochemistry and is author of *Building Biotechnology*, a widely-used course text in biotechnology programs. He regularly guest lectures for the Johns Hopkins MS/MBA program in biotechnology, teaching classes on the business of biotechnology and has published papers on diverse topics such as strategies to cope with a lack of management talent and capital when developing companies outside of established hubs, entrepreneurship in biotechnology, and new paradigms in technology-based economic development.

Yali is also managing editor of the *Journal of Commercial Biotechnology* and serves on the science advisory board of Chakra Biotech and the editorial advisory boards of the *Biotechnology Journal* and *Open Biotechnology Journal*.

Yali has a long history in biotechnology media, having created a Forbes "Best of the Web"-rated web site on the biotechnology industry for a NY Times company and managed it for many years. His other projects include the Student Guide to DNA Based Computers, sponsored by FUJI Television; BiotechBlog.com; and, DrugPatentWatch.com, a pharmaceutical industry competitive intelligence service.

Yali can be contacted at *info@thinkbiotech.com*.

Related titles from Logos Press

BUILDING BIOTECHNOLOGY
27 Figures, 14 Tables, 35 Case Studies
Second Edition
ISBN: 978-09734676-3-5

BEST PRACTICES IN BIOTECHNOLOGY EDUCATION
22 chapters on programs from 5 countries
ISBN: 978-09734676-7-3

Printed in the United States
113284LV00003B/178-231/P